Lecture Notes in Control and Information Sciences

Volume 476

Series editors

Frank Allgöwer, Stuttgart, Germany
Manfred Morari, Zürich, Switzerland

Series Advisory Board

P. Fleming, University of Sheffield, UK
P. Kokotovic, University of California, Santa Barbara, CA, USA
A. B. Kurzhanski, Moscow State University, Russia
H. Kwakernaak, University of Twente, Enschede, The Netherlands
A. Rantzer, Lund Institute of Technology, Sweden
J. N. Tsitsiklis, MIT, Cambridge, MA, USA

This series aims to report new developments in the fields of control and information sciences—quickly, informally and at a high level. The type of material considered for publication includes:

1. Preliminary drafts of monographs and advanced textbooks
2. Lectures on a new field, or presenting a new angle on a classical field
3. Research reports
4. Reports of meetings, provided they are

    (a) of exceptional interest and
    (b) devoted to a specific topic. The timeliness of subject material is very important.

More information about this series at http://www.springer.com/series/642

Harald Waschl · Ilya Kolmanovsky
Frank Willems
Editors

Control Strategies for Advanced Driver Assistance Systems and Autonomous Driving Functions

Development, Testing and Verification

Springer

Editors
Harald Waschl
Institute for Design and Control
of Mechatronical Systems
Johannes Kepler University Linz
Linz, Austria

Frank Willems
Department of Mechanical Engineering
Eindhoven University of Technology
Eindhoven, The Netherlands

Ilya Kolmanovsky
Department of Aerospace Engineering
University of Michigan
Ann Arbor, MI, USA

ISSN 0170-8643 ISSN 1610-7411 (electronic)
Lecture Notes in Control and Information Sciences
ISBN 978-3-319-91568-5 ISBN 978-3-319-91569-2 (eBook)
https://doi.org/10.1007/978-3-319-91569-2

Library of Congress Control Number: 2018941237

© Springer International Publishing AG, part of Springer Nature 2019
This work is subject to copyright. All rights are reserved by the Publisher, whether the whole or part of the material is concerned, specifically the rights of translation, reprinting, reuse of illustrations, recitation, broadcasting, reproduction on microfilms or in any other physical way, and transmission or information storage and retrieval, electronic adaptation, computer software, or by similar or dissimilar methodology now known or hereafter developed.
The use of general descriptive names, registered names, trademarks, service marks, etc. in this publication does not imply, even in the absence of a specific statement, that such names are exempt from the relevant protective laws and regulations and therefore free for general use.
The publisher, the authors and the editors are safe to assume that the advice and information in this book are believed to be true and accurate at the date of publication. Neither the publisher nor the authors or the editors give a warranty, express or implied, with respect to the material contained herein or for any errors or omissions that may have been made. The publisher remains neutral with regard to jurisdictional claims in published maps and institutional affiliations.

Printed on acid-free paper

This Springer imprint is published by the registered company Springer International Publishing AG part of Springer Nature
The registered company address is: Gewerbestrasse 11, 6330 Cham, Switzerland

Preface

Autonomous driving is expected to play an essential role in future transport systems to dramatically improve on-road safety and traffic throughput. Also, benefits are anticipated in the field of reduction of fuel consumption, of pollutant emission, and of greenhouse gas emissions. Therefore, this is seen as a crucial technology for smart, green, and integrated mobility. In this context, advanced driver assistance systems (ADASs) and automated driving functions (ADFs) can be seen as an intermediate step toward fully autonomous driving vehicles on public roads. For the development and verification of ADAS and ADF control systems, the intended use case is crucial that involves real-world driving situations with multiple participants. The control algorithms need to operate in non-deterministic and time-varying environment, which, due to sensor limitations, are not fully known and involve interacting with other traffic participants, possibly a human driver and in future even traffic management systems. As of now, demonstrators and selected functions are already available in production cars; however, still no agreement on how to design, test, and verify control strategies for these driving functions is available. The large number of combinations of possible situations render standard testing approaches, like fleet testing, virtually impossible and requires new methods and concepts to be developed. The new approaches that we need will for sure be interdisciplinary. This book aims to contribute to defining such approaches by collecting articles from experts with different backgrounds and interests to produce an overall picture of the tools we have and will need to have to cope with this problem. It covers various aspects, ranging from human interaction and theoretical approaches to real-world examples for development and testing of control algorithms for ADAS and ADF.

The starting point for this book was an international workshop on the topic held in September 2016 at the Johannes Kepler University Linz, Austria. The contents are peer-reviewed versions of selected workshop contributions. The chapters are from international experts and cover theoretical and formal methods, human interaction and behavior modeling, methods and frameworks for testing and control function development, the impact of ADF on the powertrain development and aspects of traffic control. The wide range reflects the highly interdisciplinary nature of the topic and allows a first notion on the expected challenges. For sure, it is not possible to provide all answers within one book; nonetheless, particular solutions presented to subtasks and subproblems can provide important building blocks for development, testing, and verification of control strategies for advanced driver assistance systems or automated driving functions.

Neither the workshop nor this collection of contributions would have been possible without the support of several people (in particular, Nadja Aichinger and Verena Schimpf). Further, thanks are due to the reviewers of the single chapters who have done important and essential work to ensure the quality of this volume.

Linz, Austria	Harald Waschl
Ann Arbor, USA	Ilya Kolmanovsky
Eindhoven, The Netherlands	Frank Willems

Organizing Committee

Steering Organizations

Johannes Kepler University Linz, Austria
Linz Center of Mechatronics GmbH, Austria

Hosting Organization

Johannes Kepler University Linz, Austria

Program Committee

Luigi del Re	Johannes Kepler University Linz, Austria
Ilya Kolmanovsky	University of Michigan, USA
Harald Waschl	Johannes Kepler University Linz, Austria
Frank Willems	TNO Automotive, The Netherlands and Eindhoven University of Technology, The Netherlands

Organizing Committee

Nadja Aichinger	Johannes Kepler University Linz, Austria
Verena Schimpf	Johannes Kepler University Linz, Austria
Thomas Schwarzgruber	Johannes Kepler University Linz, Austria

Referees

T. Albin	E. Atkins	A. Girard
A. Katriniok	N. Li	B. Németh
J. Ploeg	D. Reischl	R. Schmied
X. Seykens	M. Stolz	E. Tseng
C. E. Tuncali	E. van Nunen	S. Wilkins
Y. Yildiz	D. Zhao	R. Zidek

Contents

1 Cooperation and the Role of Autonomy in Automated Driving 1
Gina Wessel, Eugen Altendorf, Constanze Schreck,
Yigiterkut Canpolat and Frank Flemisch
- 1.1 Introduction 2
- 1.2 Basics of Automated Driving 3
- 1.3 Air Traffic Accidents and the First Fatal Automated Driving Crash 4
- 1.4 Cooperation and the Role of Autonomy in Different Fields 7
  - 1.4.1 The Holon—A System-Theoretical Approach 7
  - 1.4.2 Machines as Agents 8
  - 1.4.3 Responsibility: Credit and Blame in Human–Machine Interaction 10
  - 1.4.4 Conditions for (Human) Cooperation 11
  - 1.4.5 Autonomy: From Human Teams to Human–Machine Teams 12
  - 1.4.6 Cooperation and Autonomy in Vehicle Guidance and Control 13
  - 1.4.7 Summary of Different Approaches 15
- 1.5 Working Definition of Cooperation and the Role of Autonomy 15
- 1.6 Design Recommendations and Avenues for Future Research 19
- 1.7 Implementation of the Design Recommendations and Working Definitions in the H-Mode 21
- 1.8 Conclusion 23
- References 23

2 Robust Real-World Emissions by Integrated ADF and Powertrain Control Development 29
Frank Willems, Peter van Gompel, Xander Seykens and Steven Wilkins
- 2.1 Introduction 30
- 2.2 Drivers and Challenges in HD Powertrain Development 31
  - 2.2.1 Real-World Pollutant Emissions 31
  - 2.2.2 Fuel Consumption and CO_2 Emissions 32
  - 2.2.3 Powertrain Complexity 32
- 2.3 ADF Potential for Robust Powertrain Performance 33
  - 2.3.1 Integration of ADF in Powertrain Control Strategy 35
  - 2.3.2 Optimal Control Problem 36
- 2.4 Powertrain Control Development and Testing 38
  - 2.4.1 Virtual Development and Verification 38
  - 2.4.2 Laboratory Testing 40
  - 2.4.3 On-Road Vehicle Testing 41
  - 2.4.4 Potential Synergy in ADF and Powertrain Development 42
- 2.5 Conclusions 43
- References 44

3 Gaining Knowledge on Automated Driving's Safety—The Risk-Free VAAFO Tool 47
Philipp Junietz, Walther Wachenfeld, Valerij Schönemann, Kai Domhardt, Wadim Tribelhorn and Hermann Winner
- 3.1 Motivation 48
- 3.2 Virtual Assessment of Automation in Field Operation 48
  - 3.2.1 Introduction 49
  - 3.2.2 System Architecture 50
  - 3.2.3 Use Cases 51
  - 3.2.4 Correction of Environment Perception 53
  - 3.2.5 Simulation Environment 56
- 3.3 Scenario Identification 58
- 3.4 Application Ideas 60
  - 3.4.1 VAAFO Use with State-of-the-Art Vehicles 60
  - 3.4.2 Approval by "Seed Automation" 61
  - 3.4.3 Vehicle Equipment for Future Applications 62
  - 3.4.4 VAAFO on a Dedicated Fleet—A New Business Model 63
- 3.5 Conclusion 63
- References 64

4 Statistical Model Checking for Scenario-Based Verification of ADAS 67
Sebastian Gerwinn, Eike Möhlmann and Anja Sieper
- 4.1 Introduction 67
- 4.2 Running Example 69
- 4.3 Related Work 70
  - 4.3.1 Visual Logic 70
  - 4.3.2 Formal Characterization Using Stochastic Satisfiability Modulo Theory (SSMT) 73
  - 4.3.3 SSMT Solving Using Statistical Model Checking 75
- 4.4 Methods 76
  - 4.4.1 Traffic Sequence Charts 76
  - 4.4.2 Statistical Model Checking 78
- 4.5 Results 82
- 4.6 Conclusion 85
- References 86

5 Game Theory-Based Traffic Modeling for Calibration of Automated Driving Algorithms 89
Nan Li, Mengxuan Zhang, Yildiray Yildiz, Ilya Kolmanovsky and Anouck Girard
- 5.1 Introduction 89
- 5.2 Traffic Modeling Based on Level-k Game Theory 91
  - 5.2.1 Modeling of a Single Car 91
  - 5.2.2 Modeling of Interactive Traffic 95
- 5.3 Traffic Simulator Based on Level-k Game Theory 96
- 5.4 Rule-Based Automated Highway Driving Algorithm 99
  - 5.4.1 Cruise Control Mode - C 99
  - 5.4.2 Adaptive Cruise Control Mode - A 100
  - 5.4.3 Lane Change Mode - L 101
  - 5.4.4 Basic Simulation of Rule-Based Automated Highway Driving Controller 102
- 5.5 Optimal Automated Driving Controller Calibration 103
- 5.6 Summary and Concluding Remarks 105
- References 105

6 A Virtual Development and Evaluation Framework for ADAS—Case Study of a P-ACC in a Connected Environment 107
Harald Waschl, Roman Schmied, Daniel Reischl and Michael Stolz
- 6.1 Introduction 108
- 6.2 Virtual Development and Evaluation Framework for ADAS 109
  - 6.2.1 Concept 110
  - 6.2.2 Environment Synchronization—EnSim Layer 112

	6.3	Case Study Predictive Adaptive Cruise Control utilizing V2V and I2V Communication	117
		6.3.1 Control Problem Description	118
		6.3.2 V2X Based Traffic Prediction	120
		6.3.3 P-ACC Design	123
	6.4	Case Study—Performance Evaluation	125
		6.4.1 Route and Setup	125
		6.4.2 Results	126
		6.4.3 Advantages of the Proposed Framework Versus Single Tools	128
	6.5	Summary	129
	References		130
7	**A Vehicle-in-the-Loop Emulation Platform for Demonstrating Intelligent Transportation Systems**		133
	Wynita Griggs, Rodrigo Ordóñez-Hurtado, Giovanni Russo and Robert Shorten		
	7.1	Introduction	134
	7.2	State of the Art	135
		7.2.1 VIL Simulation Platforms	135
	7.3	Platform Architecture	137
		7.3.1 Intelligent Speed Recommender	137
		7.3.2 Emissions Regulation	141
		7.3.3 On the Existing Architecture	144
	7.4	New Components and Enhancements	145
		7.4.1 Scalability: Multiple Real Vehicle Embedding	145
		7.4.2 Information Exchange: Additional Sensors and Other Devices	146
		7.4.3 Map-Matching: Speed and Position Corrections	149
		7.4.4 Transmission Frequency and Code Modularity: Application Logic	150
	7.5	A Final Illustrative Use Case	150
		7.5.1 Discussion	151
	7.6	Conclusions and Future Work	152
	References		152
8	**Virtual Concept Development on the Example of a Motorway Chauffeur**		155
	G. Nestlinger, A. Rupp, P. Innerwinkler, H. Martin, M. Frischmann, J. Holzinger, G. Stabentheiner and M. Stolz		
	8.1	Introduction	155
	8.2	Requirements	156
	8.3	Software Architecture	158
	8.4	Functional Safety for Automated Driving Functions	163

	8.5	Sensor Fusion	164
	8.6	Trajectory Planning	166
	8.7	Simulation Results	169
		8.7.1 Merging Maneuver	169
		8.7.2 Collision Avoidance Maneuver	169
		8.7.3 Bumpless Transfer for Lateral Tracking	171
	8.8	Conclusion	172
	8.9	Outlook	173
	References		173
9	**Automation of Road Intersections Using Distributed Model Predictive Control**		**175**
	Alexander Katriniok, Peter Kleibaum and Martina Joševski		
	9.1	Introduction	175
	9.2	Modeling	178
		9.2.1 Vehicle Kinematics	179
		9.2.2 Inter-Vehicle Distances	180
	9.3	Centralized Conflict Resolution Problem	180
	9.4	Distributed Conflict Resolution Problem	183
		9.4.1 Primal Decomposition	183
		9.4.2 Prioritizing Coupled Safety Constraints	185
		9.4.3 Prioritization Functions	187
		9.4.4 Solving the Distributed Non-convex OCP	188
	9.5	Feasibility and Optimality	191
	9.6	Simulation Study	193
		9.6.1 Intersection Scenario	193
		9.6.2 Discussion of Results	194
	9.7	Conclusions and Future Work	197
	References		198
10	**MPDM: Multi-policy Decision-Making from Autonomous Driving to Social Robot Navigation**		**201**
	Alex G. Cunningham, Enric Galceran, Dhanvin Mehta, Gonzalo Ferrer, Ryan M. Eustice and Edwin Olson		
	10.1	Introduction	202
	10.2	Related Work	203
		10.2.1 Related Work on Behavioral Prediction	203
		10.2.2 Related Work on Decision-Making	204
		10.2.3 Related Work on Social Navigation	204
	10.3	Problem Formulation	205
		10.3.1 The Multi-Policy Approximation	207

10.4	Case Study 1: Autonomous Driving		210
	10.4.1	Behavior Anticipation	211
	10.4.2	Results	212
10.5	Case Study 2: Social Environment		214
	10.5.1	Simulation	216
	10.5.2	Real-World Experiments	217
10.6	Conclusion		219
References			219

Introduction

Advanced driver assistance systems (ADASs) and automated driving functions (ADFs) represent both important and significant steps toward fully autonomous driving vehicles on public roads. Functions, such as jam or highway pilots, offer the driver assistance in selected scenarios but do not require full control authority over the vehicle in all circumstances and conditions. An advantage of such systems is that the driver is still available as an observing and supervising entity, and functions may be deactivated if conditions for operation are not met. Nonetheless, already in these cases, for the development and verification of the respective control systems, the intended use case is crucial, which involves real-world driving with multiple participants. This means the controls algorithm need to operate in non-deterministic/time-varying and, due to sensor and system limitations, not fully known environments, while interacting with other traffic participants and possibly a human driver.

From a pure control development perspective, e.g., path planning, trajectory tracking, and powertrain control, different solutions already exist and can be found in current production or prototype vehicles. Examples are assisted driving functions intended for specific conditions or particular applications, e.g., highway driving or for operating on closed tracks like in agriculture or mining. Other examples are prototype vehicles with large sensor arrays and under supervision of skilled test drivers during the operation.

Testing and verification of control strategies for advanced driver assistance systems or automated driving functions is a highly interdisciplinary topic, which currently receives strong attention in the academic and industrial communities and is the subject of much ongoing research and development activities. These range from theoretical and formal methods to testing and verification of full systems, i.e., vehicles or even groups of them, in real-world situations. While in the past it was possible to design and verify controllers using a limited and a predefined set of test cases or in real-world fleet tests, this is not the case anymore for highly automated driving. The complex interactions between the environment, under non-deterministic and time-varying conditions, and multiple traffic participants lead to an almost infinite number of possible scenarios and situations which may be encountered and need to be handled.

As a consequence, classical approaches, such as certification and fleet testing, are not capable of handling the complexity, and so new methods to allow efficient development and testing are required. Although many academic and industrial research projects focus on this topic and aim to provide guidelines and solutions, up to now no consensus on the best practice is established. Moreover, with self-driving vehicles, new challenges arise which are not only related to technical aspects but require insights into legal, psychological, and social issues including human interaction.

It appears that simulations using models and hardware-in-the-loop facilities will be a key enabler for testing, validation, verification, and certification of the automated driving functions.

This testing in "virtual world" or mixed "real–virtual world" allows to cover the large range of potential scenarios and complexity levels involved in automated driving. To manage complexity, a modular environment which allows to simulate the system at different levels and with different real or virtual components seems to be necessary. Still, it is the question which detail levels are sufficient and can be handled. A very important component is the scenarios, i.e., the driving and environment situations, which are used in the simulation studies. These scenarios should cover all (realistic but also worst case) situations which the vehicle may encounter during the driving mission including a realistic interaction with other traffic participants and the environment. Further, it is also necessary to understand (and model) the interactions between the driver and automated driving functions, in particular, as the near-term systems will likely be supervised by the drivers and/or handover the control of the vehicle to the driver under some conditions.

During September 29–30, 2016, a workshop on the subjects of development, testing, and verification of ADAS and ADF has taken place at the Johannes Kepler University Linz to assess the status and discuss ways to improve the strategies for development and testing of such advanced driving functions. The workshop consisted of 14 talks by experts from industry and academia, covering topics in human interaction, the impact on the powertrain and emissions, testing frameworks, scenario design, and function development.

During the talks and presentation at the workshop, several interesting observations have been made related to the current status and future directions of this field. Examples of these observations are summarized below:

- Currently, in both automated driving and advanced driver assistance function development, the main focus is on improving safety and comfort of the passengers. Additional performance aspects, such as related to fuel/energy consumption, CO_2, and other tailpipe emissions, are not considered explicitly. The general assumption is that these will be dealt with later while the current focus is on vehicle trajectory and velocity planning. It is expected that once the fundamental challenges of automated driving are addressed satisfactorily, fuel economy and emission-related aspects will be fully considered. This somewhat incremental development approach may lead to automated vehicles with suboptimal overall system performance which are difficult to redesign at a later stage. Thus, pursuing holistic

opportunities involved in automated driving that involves fuel/energy efficiency and emissions appears to be worthwhile.
- There is an increasing emphasis on traffic scenario development in simulation environments. These environments seem to be a suitable tool to test and develop control functions in safe and scalable manner. It is important for the models used to be realistic including presenting responses of other vehicles, the environment and the physical properties and limitations of sensors and actuators. Depending on the intended system, test, and available time, different models and detail levels are necessary.
- The role of mixed (virtual–real) testing is also increasing, but is mainly limited to laboratory environment. There is a large potential to perform tests in the field of HIL setups with engine, powertrain, or vehicle in the loop. Modular setups which allow to exchange virtual and real components and execute the same scenario need to be developed, and one of the main challenges in this area is on a clean interface design. Here, functional mock-up units and interfaces (FMU/FMI) seem to be a valuable approach.
- Currently, a large focus is on simulations with models that are not validated from experiments, because such validations may require billions of real-world-driven kilometers. In this case, much can be gained, when the loop from the road to virtual development is closed by feeding back the field results to the virtual environments which can lead to improved uncertainty models. There seems to be a large potential for public data sharing, between OEMs, academia, and governmental agencies.
- Up to now, a standardized method for ADAS and ADF testing and certification is missing. Here, it may be important to deal with public perception and address the upcoming issues in a clear and still understandable manner. Furthermore, there are large regional differences which make the establishment of a standardized method even more difficult. Additionally, in actual testing and certification of vehicles, the impact of ADAS and ADF on CO_2 and pollutant emissions is presently not covered by legislation. Consequently, it is expected that this impact will become important for the development of future representative test cycles.
- The workshop focused mainly on passenger car applications; however, many presented methods may be applied in commercial and heavy-duty applications too.

This book combines and extends selected results presented at the workshop. It covers different aspects of development, testing, and verification of control systems for ADAS and ADF and provides insight into the underlying issues from both, theoretical and application, points of view.

Acknowledgements This work has been partly carried out at LCM GmbH as part of a K2 project. K2 projects are financed using funding from the Austrian COMET-K2 program. The COMET-K2 projects at LCM are supported by the Austrian Federal Government, the Federal State of Upper Austria, the Johannes Kepler University, and all of the scientific partners which form part of the COMET-K2 Consortium.

Harald Waschl
Ilya Kolmanovsky
Frank Willems

Chapter 1
Cooperation and the Role of Autonomy in Automated Driving

Gina Wessel, Eugen Altendorf, Constanze Schreck, Yigiterkut Canpolat and Frank Flemisch

Abstract While automation in human–machine systems can increase safety and comfort, the 2016 lethal crash of an automated vehicle demonstrates that automation is not without its risk. Indeed, accidents from the air traffic domain also demonstrate that cooperation between human and machine is crucial and interaction design must be devoted great care. Therefore, the present paper aims at developing design recommendations to reduce the risks of lethal crashes with automated vehicles. To this end, the concepts of cooperation and autonomy are closely investigated. These two terms are central to research on human machine cooperation; however, the present definition of cooperation and the role of autonomy might be further specified for the domain of automated driving. Therefore, selected perspectives from different scientific fields (e.g., sociology and psychology) will be presented in order to develop a differentially inspired working definition of cooperation, which is tailored to the automated driving domain. Another goal of this approach is to investigate different views on the concept of autonomy, which is often entailed in work on cooperation. This can help clarify the role of autonomy in automated driving in particular. Moreover, insights from the presented theories and findings on cooperation can be transferred to the interaction design of automated vehicles. Accordingly, recommendations for interaction design will be presented. Finally, an example for the implementation of the working definition and the design recommendations will be presented by describing a prototype for automated driving—the H-mode prototype.

G. Wessel (✉) · E. Altendorf · C. Schreck · Y. Canpolat
Institute of Industrial Engineering and Ergonomics,
RWTH Aachen University, Aachen, Germany
e-mail: g.wessel@iaw.rwth-aachen.de

F. Flemisch
Fraunhofer Institute for Communication, Information Processing and Ergonomics FKIE,
Wachtberg, Germany

1.1 Introduction

> "Alone we can do so little; together we can do so much"
>
> Helen Keller

This quote of a deaf-blind author seems strikingly simple and true, yet the application of its meaning is not that easy. Cooperation between humans can be a complex issue, and cooperation between humans and machines might be even more intricate. Which roles do human and machine play? Who is responsible for joint actions? How much autonomy should a machine have? Or how much autonomy of a machine is needed before cooperation can take place? And is cooperation with a machine even possible at all?

A special field of application of these questions is the domain of automated driving. In this domain, the concept of highly automated and cooperative automation prevails and emphasis is put on the advantages of combining the assets of driver and automation for creating a high-performing system, e.g., [36, 38]. The term cooperative guidance and control is used to denote a focus on these two aspects of driving. In the framework of these concepts, driver and automation need to act jointly, which makes questions about the automation's autonomy inevitable [6, 21, 22, 25].

Even though the concepts of cooperation and autonomy have been defined in a general way by dictionaries, their range of application is very wide. Therefore, several domains have specified and defined both concepts in a tailored way in order to organize research, for example, in teamwork [31] or biology research [32]. For the automated driving domain, the following definition of cooperative guidance and control has been presented:

> the action or process of working together of at least one human and at least one computer on the guidance and control of at least one vehicle [25].

Furthermore, Flemisch et al. [21] claim that at higher levels of automated driving the capability of the machine to function autonomously is a necessary prerequisite for cooperative driving. Additionally, they point out that at higher levels of autonomy, there might be a tradeoff with cooperativeness. These two statements demonstrate the interdependence of the two concepts and point out the need to provide further explanation.

While this shows that these concepts already received attention and tailoring to the domain of automated driving, this domain is still a relatively new scientific field. Therefore, definitions and research findings concerning cooperation and the role of autonomy from different domains could inspire formulating an elaborated working definition of cooperation and clarify the role of machine autonomy in driving. Furthermore, a solid foundation of the concepts of cooperation and autonomy could help develop recommendations for interaction design. This is of major importance, because some crashes with automated planes and the 2016 fatal Tesla crash can (partly) be ascribed to bad interaction design.

Accordingly, in the following, general concepts and frameworks from the domain of automated driving will be provided and an airplane crash and the 2016 Tesla vehicle crash will be described as an example. Selected perspectives, theories, and research from different fields—including but not limited to the domain of automated driving—on the topics of cooperation and autonomy will be presented. The concepts from other domains are then transferred to the domain of highly automated and cooperative vehicle guidance and control in order to reach a tailored working definition of cooperation, to elaborate on the role of autonomy and derive recommendations for system—and interaction design. Lastly, an implementation of the derived working definition and design recommendations is presented by the example of a prototype for a cooperative automation: the H-mode prototype.

1.2 Basics of Automated Driving

Equipping ground vehicles with autonomous abilities has been a research topic for four decades. Today, first systems that allow automated driving are already in serial production and available for consumers. Nonetheless, in currently available automated vehicles, a human driver is still present and expected to overrule the technical system in any case of dysfunction. Moreover, in complex situations, which are not solvable for a rule-based technical system, the human driver with her ability for creating new solutions to unforeseen problems is indispensable. In the following, concepts and frameworks that are relevant to the discussion about autonomy and cooperation in automated driving will be introduced.

When humans are working together with machines, both entities hold varying degrees of control. This can be depicted as a continuum ranging from completely manual control to full automation control—in this case from manual to completely autonomous driving, see Fig. 1.1 and [24, 26, 34].

In line with this notion of a control distribution, different

> levels of automation of decision and action selection

have been defined by Parasuraman et al. [51]. These range from the lowest level of automation (Level 1), which denotes completely manual control, to the highest level of automation (Level 10) denoting complete machine control. Based on the notion of a control distribution, different institutions, e.g., Society of Automotive Engineers International [6] or the German Federal Highway Research Institute (BASt) [29], have defined discrete levels of automation for automated driving (see Fig. 1.1). The term autonomous driving has been used inconsistently—either for all levels of automated driving, for all levels above Level 3 or for full automation. Therefore, SAE [6] recommends completely avoiding this term for describing automation capacity. Currently, implementation up to SAE Level 2 is commercially available for customers, while fully automated vehicles are not yet feasible. Accordingly, the interaction design of automated vehicles is a key factor for currently available systems. This is demonstrated by the following examples of aircraft accidents and the 2016 Tesla vehicle crash.

	0	1	2	3	4	5
SAE	No Automation	Driver Assistance	Partial Automation	Conditional Automation	High Automation	Full Automation
BASt	Driver Only	Assisted	Partially Automated	Highly Automated	Fully Automated	---
NHTSA	0 No Automation	1 Function-specific Automation	2 Combined Function Automation	3 Limited Self-Driving Automation	3/4 Full Self-Driving Automation	

Fig. 1.1 Comparison of SAE, BASt, and NHTSA taxonomies [5, 6, 20, 24, 29, 46]

1.3 Air Traffic Accidents and the First Fatal Automated Driving Crash

As automated driving is only recently becoming available, accidents due to automation deficits or complications concerning the interaction between human and automation have been rare. Opposed to this, in the air traffic domain several plane crashes could be seen as an example for interaction intricacies between human and automation. Even though traveling by plane is one of the safest traveling options [63], the long history of flight deck automation also leads to a history with more fatal accidents [39]. An example is the crash of the Aeroflot Flight 593 in 1994, which caused the death of all 75 persons on board. The pilot of the Airbus 310-304 turned on the autopilot and let his children play with the control wheel. His son's steering actions interfered with some functions of the autopilot, which was recognized too late, ultimately leading to the crash [1]. Another example is the crash of France Flight 447 in 2009. The autopilot of the Airbus A330-203 was deactivated due to icing issues, and ultimately the crew was unable to control the plane in manual mode [11]. As a reaction to accidents like these, several different measures have been realized. For example, glass cockpits, crew resource management, and improvements of the automation and the HMI were introduced as results of extensive analyses of the causes of air craft accidents [60, 62, 72].

A major problem that has been addressed with these measures is mode confusion. Butler et al. [13] mention three possible sources of mode confusion that derive from automation and user characteristics: opacity, complexity, and incorrect mental models [13]. These issues can lead to an incorrect understanding of the state the automation actually resides in. Therefore, mode confusion is closely linked to a loss of situation awareness (even though they are generally addressed as different concepts because of different emphases): Situation awareness (SA) denotes

> being aware of what is happening around you and understanding what that information means to you now and in the future [17].

For that reason, the actual mode of an automation has to be transparent and comprehensible in order to ensure SA and to prevent mode confusion [17]. As there clearly are parallels between (problems with) interaction design in air and ground vehicles, an exemplary plane crash, which was linked to mode confusion, will be presented in detail for illustrative purposes.

The Everglades plane crash happened on December 29 in 1972, where a plane (Lockheed L-1011 TriStar) of Eastern Air Lines crashed into the Florida Everglades while operating the flight from John F. Kennedy International Airport in New York City to Miami International Airport, Florida (Flight Number 401). The flight started to approach Miami Airport, when the indicator of the nose landing gear failed to light up after lowering the gear. Therefore, the pilots started to look for problems related to the nose gear and tried to get the confirmation of the indicator. During this time, the pilots set the autopilot of the plane to keep the flight level on 2000 ft (610 m). But the plane left this flight level after a while unintentionally and began a gradual descent. The pilots were completely distracted by the nose landing gear problem and realized the descent too late. An audible alert, which signaled the unplanned descent before getting into the safety critical situation was not registered. Only seconds before the crash, pilots tried unsuccessfully to save the plane. This resulted in the death of 99 persons.

The investigations of authorities suggested that the pilot had accidentally initiated the descent as he leaned against the control stick while he was turning to speak to the flight engineer, who was sitting behind him. The slight forward pressure on the stick would have caused the autopilot to enter a descent. Furthermore, the distraction of the pilots due to the problematic nose landing gear hindered the operators to recognize the danger in time [47].

Conclusively, this accident illustrates the concept of mode confusion. While the crew thought that the aircraft was set on autopilot to keep the flight level at 2000 ft, it was in descending mode. Furthermore, this accident indicated the need for better crew resource management, assuring that not all crew members would focus on a single (rather unimportant) problem. As mentioned above, different measures have been taken in the air traffic domain to prohibit further accidents like this one.

A tragic accident in the automotive domain, which was also partly due to missing cooperation between automation and operator, is the 2016 Tesla crash. In this fatal crash on May 7 in 2016 at 3:40 p.m., 45-year-old Joshua Brown died after his Tesla Model S collided with an 18-wheel semi-truck in Williston, Florida. At the time of the accident, the autopilot system of the vehicle was switched on, which enabled control of the longitudinal and lateral dynamics of the vehicle by an automation system of the vehicle according to SAE Level 2 (partial automation). The automation did not stop for the truck, which was perpendicularly crossing the divided highway with a median strip, by a left turn onto a side road. The vehicle collided with the truck at

74 miles per hour (65 miles per hour was allowed according to traffic regulation) without the driver or the autopilot system ever applying the brakes. The Tesla vehicle subsequently passed under the center of the trailer and stopped at the side of the road approximately 100 ft south of the highway after hitting a fence and a pole.

The Tesla autopilot system described above uses cameras and a radar system to detect and avoid obstacles. However, the cameras were unable to differentiate between the white side of the trailer and a bright sky. The radar system detected the trailer, but the implemented radar parameters lead to a misinterpretation of the radar data for the automation, which recognized the trailer as an overhead road sign due to its height and its perpendicular position to the road.

As drivers activate the described Tesla autopilot system, the acknowledgment box in the HMI of the vehicle explains that the autopilot system

> [...] is an assist feature that requires you to keep your hands on the steering wheel at all times [...],

and that

> [...] you need to maintain control and responsibility for your vehicle [...]

while using it [2]. Additionally, the automation reminds the driver to

> [...] be prepared to take over at any time [...],

while sensors assess whether the hands remain on the steering wheel. When the hands are removed, an additional, visual and auditory warning is provided [2]. Furthermore, the vehicle slows down gradually if the warnings are not successful and the driver does not put his hands back on the steering wheel. However, the driver of the Tesla vehicle in the fatal accident did not show any reaction, which implies that he was probably distracted by a non-driving-related task and was not monitoring the ongoing traffic. The Florida Highway Patrol confirmed that a portable DVD player was found in the wreckage, which indicates that he might have been distracted by watching a movie [2].

An additional tragic aspect of this crash is that Joshua Brown, as a former active member of the Tesla community, described his doubts and the potential dangers of the autopilot system in an earlier comment to an uploaded video of a ride with his vehicle. Thus, he was aware of the potential dangers in using the autopilot system, but nevertheless was unable to protect himself from these dangers. Obviously, it is necessary to better adapt automation design in these early stages of ground vehicle automation to the human user with his strengths and limitations. This crash indicates that issues concerning overtrust and related phenomena need to be taken seriously and cannot easily be countered by awareness of these problems. This has also been demonstrated in scientific research before [9, 71].

The 2016 Tesla crash and the Everglade plane crash, though having different causes, both demonstrate how important it is to design an automation system, which supports cooperation between driver and automation for avoiding any kind of safety critical situations. Several approaches that have been used in the air traffic domain should be considered for automated driving as well. Especially in the present time, where autonomous driving is not yet feasible, cooperation and the role of autonomy constitute a major issue for the driving sector. Theories, research, and insights from different scientific fields will be provided in order to help preventing future crashes.

1.4 Cooperation and the Role of Autonomy in Different Fields

There is a multitude of different scientific perspectives on the concepts of cooperation and autonomy. Examples are conceptualizations of cooperation and cooperativeness from philosophy (e.g., [15]), sociology (e.g.,[64]), chemistry (e.g., [12]), science and technology studies (e.g., [55, 57]), political science (e.g., [61]), biology (e.g., [32]), and linguistics (e.g., [18]). In almost all of these cases, different conceptualizations of cooperation exist and frequently questions are raised about autonomy. If a machine can be seen as a potential cooperation partner, findings from these different fields can be compared and applied to the context of human–machine interaction in automated driving.

Therefore, the following subsections focus on an excerpt of theoretical approaches, which have been selected based on an extensive literature research on the keywords "cooperation," "cooperativeness," "collaboration," "control," and "autonomy". The most transferable approaches have been selected for presentation. This selection is not meant to be exhaustive, but can only give an overview of other application domains of the concepts in order to inspire a working definition of cooperation, which meets the needs of automated driving and to clarify the role of autonomy in this domain.

While all the following subsections target the concepts of cooperation and autonomy, the first two subsections also support the general claim that a vehicle's automation can be seen as a cooperation partner. The remaining three subsections present research, which is suitable for deriving information on interaction design for driver–vehicle cooperation.

1.4.1 The Holon—A System-Theoretical Approach

In system theory, the concepts of autonomy and cooperation play a central role. In a general definition, a system can be seen as a

> [...] deliberate arrangement of parts (e.g., components, people, functions, subsystems) that are instrumental in achieving specified and required goals [37].

A system consists of elements which have certain qualities and functions whereas each element can be regarded as an individual system as well. These sub-systems are connected through relationships and can again be further distinguished into (sub-)sub-systems (e.g., [30]). Usually, system boundaries can be defined as being context-specific to enable suitable analysis. This means that the definition of a system is not ultimate, but defined by its analysis or modeling [53]. When applied to automated driving, the driver and the automation can be seen as two different entities—each with its own capabilities, activities, and also autonomy. On the other hand, they are both integrated sub-systems in the driving context as a whole. This system-theoretical approach can also be backed by Koestler's holon concept, which has already been applied to technological systems, e.g., to manufacturing systems (e.g., [14, 33]). The basic idea of this approach is:

> A part is whole is a part. Each sub-whole is both a sub and a whole [44].

As a whole, the sub-whole has the competence to act autonomously and spontaneously and is

> [...] modified but not created by the environment's input [...] [44].

The essential message of the holon approach is that elements (holons) have both: autonomy and cooperativeness [33]. Babiceanu and Chen [8] describe a holon in the manufacturing approach as commonly consisting of hardware and software and sometimes even of software alone. Accordingly, a vehicles automation can be seen as a holon. This in turn implies that according to the holonic approach it would possess the competences to cooperate and to act autonomously. Babiceanu and Chen [8] provide working definitions of several concepts tied to the holonic approach based on Christensen [14], which can explain the connotation of these terms in the present context. They state that **cooperation is**

> [...] **a process whereby a set of entities develops mutually acceptable plans and executes these plans.**

And autonomy denotes

> [...] **the capability of an entity to create and control the execution of its plans and/or strategies** [8].

According to these working definitions, an automation is an autonomous holon, which can cooperate with a human partner.

1.4.2 Machines as Agents

In line with the holonic approach in which a machine and a human as well as the human–machine system can be seen as a holon, Rammert views machines as possessing agency (e.g., [54–57]). While traditionally machines have been seen as functioning and humans as acting entities, Rammert developed an approach for overcoming this dualism of human action and the pure functioning of machines. Instead of being simple tools with completely predictable functions, he sketches technology as possessing agency. This viewpoint is supported by different lines of research, which directly focus on human–machine cooperation [35, 49]. Rammert justifies this position by explaining that different technologies are built up from several sophisticated part systems, gain more and more autonomy, can act goal-directed, and react differently than actually was intended by the user due to own assessment of the situation. This also enables real communication between operator and machine as it exceeds simple command (or question) and response interaction [55]. Though no clear definition of autonomy is presented by Rammert and Schulz-Schaeffer [57], they argue that even humans are not completely autonomous, because factors like social interaction and social structure lead to non-intended results of human action.

Flemisch et al. [22] elaborate on this point. According to them, the autonomy of any agent is restricted due to disturbances, which influence the system's behavior in an unplanned and unwanted manner. Thus, any agent is not able to execute his task and to control the system behavior as intended. As these disturbances are more or less unpredictable, according to this view, being completely autonomous is unachievable for technical agents.

This is compatible with Kaber's work on machine autonomy [41]. He points out that autonomy is a multi-faceted concept, which consists of viability, independence, and self-governance of the machine. He emphasizes that an autonomous machine needs to be able to adapt on its own without any support from the environment, human, or task protocols. Therefore, he refuses levels of autonomy and sketches the concept as an ideal state, in which all requirements are fulfilled. From the arguments presented above it could be concluded that **machines never reach this ideal state of autonomy. However, opposing to Kaber [41], Rammert presents autonomy as a continuum, so that machines can gain more and more autonomy. For him the term agency seems to denote a critical point of autonomy on this continuum, which can be reached by machines** [57].

Based on these capabilities, possibilities, and the reflexivity of technology, Rammert and Schulz-Schaeffer [57] show that restricting the area of actions exclusively for humans is inadmissible. However, they argue that the future of a wide range of technologies (e.g., autopilot systems) does not lie in completely autonomously acting machines, but in cooperating and adapting to the user, here: the driver. **No clear definition of cooperation is given by Rammert [54] or Rammert and Schulz-Schaeffer [57], but it becomes clear that striving for a common goal seems to be a part of cooperation as well as the interaction of entities.**

Furthermore, this perspective focuses on heterogeneous systems in contrast to functional separation of tasks and sub-systems. Distribution of action between user and machine has to be taken into account right from the start. Technological systems are gaining more and more agency in complex tasks, and contingency is increasing. They are no more passive instruments but active agents. For these reasons, Rammert and Schulz-Schaeffer [57] demand a new perspective on technology and action.

In line with that, Friedman [28] demonstrated that most people feel that computers have decision-making capabilities and approximately 50% judge computers to have intentions. While the truth of this assessment is heavily debated (e.g., [16, 27, 40]), it signals that users might indeed see computers or machines as actors and therefore as potential interaction partners.

Even though there are different standpoints concerning agency of computers and machines, there are perspectives on technology, which allow the transfer of rules, strategies, habits, etc., of cooperation between humans to the cooperation between humans and machines. When applying the theories and findings presented above to human–machine interaction in automated driving, an automation can be seen as an agent, and conclusively, concepts of communication and cooperation between two agents can be transferred. Furthermore, the work presented above denotes that autonomy might either be seen as a continuum, on which different, discrete states could be determined or as a discrete, ideal state.

1.4.3 Responsibility: Credit and Blame in Human–Machine Interaction

From the two subsections above, it becomes clear that a machine and therefore also a vehicle automation can be seen as a contingent agent. However, when contingent agents are working together as a socio-technical system, questions about responsibility of both partners arise. In this subsection, it is important to mind the distinction between objective debates on how much responsibility a machine/computer can objectively be ascribed based on technological and philosophical grounds and the responsibility users actually ascribe them. Concerning the former, currently, the driver is always responsible for the vehicle's actions. This is due to current legislation, as the available implementations of automated driving do not exceed Level 2 (partial automation) according to SAE standard J3016 [6], respectively the highly automated level according to BASt [29]. Therefore, to date, according to the revision of the Vienna convention on road traffic, the driver always has to be in "supervisory control" [68]. In line with this, the present paragraph will focus on how much responsibility users actually ascribe to machines.

This line of research focuses on collaboration rather than cooperation. E.g., Kim and Hinds [42] investigated human–robot collaboration. **No clear definition of collaboration is given, but working together seems to be a crucial part of it [42]. Accordingly, the term strongly resembles cooperation or seems to have considerable overlap.** Since Flemisch et al. [21] claim that in human–machine interaction both terms can be treated similarly, results from the present research could be informative for driver–vehicle cooperation.

Friedman [28] previously conducted research in this field and demonstrated that humans attribute agency and moral responsibility to computers. Later, Kim and Hinds [42] investigated the question of how autonomy and transparency of behavior of a service robot influence user's attributions of credit and blame. Transparency denotes explanations of own behavior offered by a machine. The authors define two levels of autonomy, namely

> **(1) high autonomy with little need of human intervention and (2) low autonomy with need of constant human intervention** [42].

Their results show that more autonomous robots are blamed more, but do not receive more credit. This finding implies that humans might use the robot as a scapegoat. While this might in itself not be problematic, the authors acknowledge that this could reduce the operator's conscientiousness in a task. Furthermore, it was found that transparency can decrease attribution of blame toward other co-workers, but not toward the robot [42]. This supports the possibility of inducing lowered conscientiousness.

Under the term complacency bias, a related phenomenon has been documented. Complacency bias denotes

> a psychological state characterized by a low index of suspicion [71],

which has been documented by studies investigating humans working with automated machines and can lead to errors and lower system functioning [50]. In the context of automated driving, these findings can hint potential problems. Due to current legislation the driver is always responsible for the vehicle's actions [68], but complacency bias could lead to accidents. A tendency to blame the machine for occurring problems might even increase likelihood for that.

The above-presented research show that users might in some cases overestimate an automation's capacities and feel less responsible themselves. Even though legislation is different, designers of automated vehicles might want to keep this potentially dangerous human tendency in mind.

1.4.4 Conditions for (Human) Cooperation

The above-presented tendency to blame others might be seen as a common and natural human phenomenon [66, 67, 70], which should be kept in mind when designing an automation. This calls for the question whether other aspects of human social behavior could inform the debate on cooperation and autonomy. To this end, a biological and sociological approach to cooperation will be presented.

Hamilton and Axelrod [32] (the first is an evolutionary biologist, the second a political scientist) outline the evolutionary process of living organisms from the level of microbes on by focusing on cooperation. The game theoretical prisoners dilemma forms the basis of this approach, and one of the key messages of their research is that cooperation is based on reciprocity. Later, Axelrod [7] extends this approach and identifies three main conditions to promote (human) cooperation: The first condition entails a likelihood of meeting in the future. This is necessary to provoke accountability for our actions. However, as Axelrod states, there is no need of foresight but it can promote cooperation. The second condition consists of an ability to identify each other: Actors have to identify whom they are coping with in order to recognize the other as a sane counterpart. Knowing a few characteristics of the other is useful for estimating the other's cooperativeness and strategies before cooperation starts. The last condition entails a record of past behavior: Only when past experiences are connoted in a positive way, there will be future cooperation. Here information about reputation is faced, which can again give hints on cooperativeness of the other. While these conditions promote cooperativeness, there can be cooperation without fulfilling all of them. Furthermore, these assumptions show that questions about past and future are essential to clarify processes and actions in the current situation [7].

When cooperation should be enhanced in the domain of cooperative driving, these conditions can be used as a guideline as will be outlined in the section on design recommendations below.

1.4.5 Autonomy: From Human Teams to Human–Machine Teams

Another domain, from which the design of the cooperation between driver and automated vehicle might profit, is research on teamwork. According to the definitions presented above, machines could be seen as possessing autonomy. However, more conservative definitions of autonomy have been formulated in the domain of work psychology. For example, Hackman and Oldham [31] investigate employee motivation in workplace contexts and identify autonomy as a critical variable. They define **autonomy as the**

> **degree to which the job provides substantial freedom, independence, and discretion to the individual in scheduling the work and in determining the procedures to be used in carrying it out** [31].

While this description focuses on autonomy as an objective condition, Kirkman et al. [43] describe autonomy in the context of teamwork as a subjective experience, namely

> **the degree to which team members believe they have freedom to make decisions** [43].

According to this definition, it is questionable whether an automation can ever be autonomous. The definition by Kirkman et al. includes a subjective evaluation of personal freedom, which (at least for the present time) can be precluded. However, according to the definition by Hackman and Oldham [31], autonomy could be applicable to an automation's "work," because in autonomous driving it could be given freedom, independence, and discretion in scheduling and choosing procedures. This is illustrated by the following example: In the future, a completely self-driving car might have the freedom to make changes to the route in order to prevent future traffic jam without consulting the passengers. Accordingly, it is dependent on the scope and strictness of the definition of autonomy whether a vehicle's automation can be seen as being autonomous.

Even though autonomy according to some work psychological definitions might not be ascribed to an automation, as it has no needs and no subjective experience, these approaches could be valuable for informing human needs in cooperation. Indeed, in the field of automated driving it might be necessary to provide different definitions of autonomy for human and machine. Alternatively, different degrees of autonomy could be formulated and different target levels could be handled for each partner. As Hackman and Oldham [31] claim, autonomy according to this definition is important for human motivation. A machine is not dependent on motivational factors, but for humans motivation might be an important factor for cooperation. However, before a working definition on cooperation and the role of autonomy in the framework of driver–vehicle cooperation can be drawn, previous work from this domain needs to be investigated.

1.4.6 Cooperation and Autonomy in Vehicle Guidance and Control

In the following, the literature on cooperative driving and the role of autonomy in cooperative driving will be presented. Flemisch et al. [25] present a list of aspects, which can contribute to the cooperation between driver and automation. Among others they name

> traceability and predictability of abilities and intents in both directions

and

> sufficiently autonomous machine capabilities for higher levels of automation of guidance and control [25].

Cooperation between the automation and the human driver in vehicle guidance and control can be on the vehicle's lateral and/or longitudinal dynamics (e.g., on the longitudinal dynamics with an Adaptive Cruise Control). According to Flemisch et al. [19], cooperation on the control level is described as shared control, which is also included in the concept of cooperative guidance and control of vehicles. Thus, sharing the control of minimum one dimension of the vehicle dynamics can be considered as cooperation in vehicle guidance and control. Nevertheless, both dimensions are necessary for full vehicle control and they cannot be seen as being completely separated, as they exert an influence on each other. If control authority should be given to the automation, it must be assured that the automation has a principle **understanding of lateral and longitudinal control, their interactions, the situation, and the human's (probable) actions** in order to predict the outcomes of actions and provide support [21, 25, 58]. This is in line with Pollard et al. [52], who propose an ontology for automation levels in automated driving. Ontology in this context denotes a representation of knowledge inside the automation. A vehicle with different levels of automation has to assess the situation including its own capabilities and knowledge. Hence, some kind of self-reflection is required:

> To guaranty safety, it is absolutely necessary to assess what the vehicle knows and to adapt the driving behavior to the perception uncertainties [52].

Accordingly, the ability to communicate is needed for being cooperative [52].

In addition to that, Hoc [35] and Hoc et al. [36] claim that cooperation commonly entails interference, which denotes an (inter-) dependency of goals of both cooperation partners. This assumption highlights that both, human and automation, have their own goals concerning the driving task, which leads to a need of managing interference on different levels. Thereby, the authors' primary focus lies at dynamic situations where high amounts of uncertainty occur. These dynamic and uncertain situations need to be addressed by cooperation, because both partners have special capabilities that are required to handle these situations [35, 36].

While the present chapter focuses on driver–vehicle cooperation (vertical), a vast amount of literature on cooperative driving focuses on vehicle–vehicle cooperation (horizontal). For example, Shladover [65] sees cooperation in the context of platooning automated vehicles as the inverse of autonomy. That is because he defines **autonomous vehicles as**

> **depending entirely on their own sensors as information sources, without communication to or from other vehicles or the infrastructure** [65].

This is in line with SAE standard J3016 [6], in which it is stated that some vehicle automations might be seen as autonomous if they do not depend on other agents in any of their functions. However, if they depend on other entities, they are defined as being cooperative rather than autonomous. Consequently, cooperation eliminates autonomy according to SAE [6].

Opposed to that, Rödel et al. [59] claim that autonomy needs to be granted for an ideal interaction. If there is sufficient autonomy of the automation, the driver can give away control in certain situations. To this end, a certain degree of trust in the vehicle (automation) is required [59]. This is in accordance with their definition of **autonomy as constituting the**

> **ability and authority to make decisions independently and self-sufficiently** [59].

In addition, Vanderhaegen [69] defines characteristics for achieving autonomy for general agents such as a human or a machine. These characteristics, e.g., include having a set of behaviors to choose from, being capable to interact with other agents, the ability of independently choosing the own actions, and the capability of striving for and achieving goals. Furthermore, the requirements need to be met to go through with the present goal without the help of additional external resources.

Thus, even though different views on the term autonomy exist in the literature, most authors agree on autonomy as being a key component for an intelligent agent in vehicle guidance. It can therefore be stated that some degree of autonomy seems to be necessary for cooperative vehicle control.

1.4.7 Summary of Different Approaches

In the previous sections, different conceptualizations of cooperation and autonomy and theories and findings from different fields of research have been presented. While this is only a selective excerpt from existing research, several implications for the automated vehicle domain can be drawn. Firstly, it is important to determine the focus of attention: Is the holon a driver–vehicle system or is it either driver or vehicle? Both views might be informative, but for the present context the automation as a holon is the most relevant aspect. View as a holon, the automation possesses autonomy and cooperativeness.

This view of an automation as an actor is shared by Rammert and Schulz-Schaeffer [57], who claim that machines possess agency. Even though there are different views on that, the existence of these viewpoints is encouraging enough for attempting to implement insights from human–human cooperation to driver–vehicle cooperation. Furthermore, empirical results indicate that users might see computers as actors. This can be seen as being even more important than the objective discussion, because increasing safety and comfort for the user and other persons involved is the main goal of cooperative guidance and control [25]. This leads to the possibility to extract design recommendations from different fields of research.

On the one hand, research in the field of perceived responsibility and the according behavior depicts potential dangers, which should be kept in mind. On the other hand, Axelrod's conditions to promote (human) cooperation could be implemented in the development of future vehicle automation systems. Still other research—like the presented research on teamwork—seems to point in the direction that different conceptualizations of autonomy for driver and automation might be useful. Directly from the automated vehicle domain, the SAE [6] described autonomy as a combination of ability and authority to make decisions. In the following, a working definition on cooperation and autonomy will be presented based on the theories, findings, and considerations from the presented fields and the current research in the domain of automated driving.

1.5 Working Definition of Cooperation and the Role of Autonomy

On the one hand, Babiceanu and Chen [8] defined cooperation based on Christensen [14] as

> a process whereby a set of entities develops mutually acceptable plans and executes these plans.

On the other hand, Rammert [54, 55] and Rammert and Schulz-Schaeffer [57] stress the interaction of entities and the existence of a common goal as parts of cooperation. This is also in line with one of the Oxford dictionary definitions of the action or process of working together to the same end [48].

Flemisch et al. [25] already tailored this definition to the context of automated driving, which they called cooperative guidance and control and defined it as follows:

> the action or process of working together of at least one human and at least one computer on the guidance and control of at least one vehicle.

However, based on the notions elaborated above, the definition by Flemisch et al. [25] could be refined by specifying the common goal(s) the partners are working toward as well as integrating the aspect of making and executing plans: Reaching a destination seems to be a plausible goal for cooperative guidance and control. Though it might be argued that mere pleasure could be the goal of driving in certain circumstances, this goal will coexist with either the goal of finally getting back to the point of departure or reaching another destination. Furthermore, the goal of safely reaching a destination should be a common goal, even though it is possible to construct situations in which it is not, e.g in military operations or in the extreme Kamikaze missions. However, these situations are not the focus of cooperative driving, which is concerned with increasing safety and comfort [25]. Therefore, the safety aspect could be included as a goal in the working definition.

Other possible goals that could be shared are reaching the destination fast or comfortably, however these can be seen as sub-goals to the goal of reaching the destination safely. This leads to the claim of integrating the aspect of having the option to make mutually acceptable plans and executing these. Ideally, an automation should be constructed in a way that is adaptive to the driver and her preferences, e.g., concerning preferred speed, safety distances, or lane positioning of the vehicle. However, for cooperation it is necessary that the automation ensures agreement of the driver before, e.g., through the driver allowing the system to overtake by itself in advance or through the system communicating the plan to the driver before executing it. This agreement of the driver might even be given just by choosing the automation mode. Based on these considerations, the following working definition of cooperative driving is provided:

In the domain of automated driving, cooperation between driver and automation denotes the action or process of working together of at least one human and at least one computer on the guidance and control of at least one vehicle with at least the one common goal of the agents of reaching a destination safely. In order to reach these goals, the agents can make plans concerning vehicle guidance and control, which are mutually agreed on, and execute these plans.

This cooperation can only be accomplished if the automation has a certain degree of autonomy: Table 1.1 depicts the different conceptualizations of the term autonomy that have been introduced above. It is interesting to note that some authors see autonomy rather as ability (e.g., [8]), while others conceptualize it as freedom or independence (e.g., [31, 43]), while the SAE [6] describes it as both.

In line with these conceptualizations, here, autonomy will be conceptualized as a meta-ability to act and control a situation without the support of other agents. A certain level of abilities is needed for autonomy, but autonomy itself is not simply one

1 Cooperation and the Role of Autonomy in Automated Driving

Table 1.1 Different definitions and conceptualizations of autonomy

Author	Context	Conceptualization/definition autonomy
Babiceanu and Chen (2006) after Christensen (1994)	Holonic approach	The capability of an entity to create and control the execution of its plans and/or strategies
Kim and Hinds (2006, p. 2)	Research on the effects of transparency and autonomy of robots	Operational definition: "high autonomy with little need of human intervention" "low autonomy with need of constant human intervention"
Rammert and Schulz-Schaeffer (2002) Rammert (2002, 2003, 2006) and	Sociological approach of machines as agents	No clear definition of autonomy is presented by Rammert and Schulz-Schaeffer (2002), but autonomy is presented as an idealistic state that machines can never reach. However, it is also conceptualized as a continuum, so that machines can gain more and more autonomy. For Rammert and Schulz-Schaeffer (2002), the term agency seems to denote a critical point of autonomy, which according to them can be reached by machines.
Kaber (2016)	Difference between automation and autonomy	Conceptualization of autonomy as a discrete ideal state, characterized by viability, independence, and self-governance.
Hackman and Oldham (1976)	Work Psychology	"Degree to which the job provides substantial freedom, independence, and discretion to the individual in scheduling the work and in determining the procedures to be used in carrying it out".
SAE International (2016)	Cooperative vehicle guidance and control	Autonomy can be seen as "the ability and authority to make decisions independently and self-sufficiently".

of these abilities at the same level, e.g., the ability for longitudinal control. With this conceptualization, it can be distinguished between potential autonomy — a meta-ability which could be used — and realized autonomy — when this potential is being exploited. This differentiation between potential and realized autonomy enables a consolidation of both types of definition. In the driving domain, the different modes of driving [6, 29] denote how much autonomy needs to be realized. However, the capability to perform certain acts is defined by the degree of autonomy, which is among others determined by the lower-level abilities and the authority to act.

Another aspect that varies between the definitions of autonomy presented in Table 1.1 is their strictness. Some authors seem to leave true autonomy to be reserved for humans (e.g., [43]). While this research is well suitable for deriving design recommendations from minding the influences of autonomy on human motivation. However, in the following it is focused on autonomous capacities of the automation, as this is in line with the overall focus of this chapter and definitions from other authors allow (some degree of) autonomy of a machine [8, 31, 42, 55, 57]. Some authors hint that there is no single definition of autonomy, but that autonomy might be seen as a continuum (e.g., [55]), while others present it as an ideal state [41]. As in the present approach autonomy is seen as a meta-ability, the former perspective is chosen for practical reasons. The levels of automation defined by SAE and BASt [6, 29] demonstrate that there is a continuum of control between driver and automation. It seems that higher levels of automation presume more autonomous capacities of the automation. Therefore, a perspective of autonomy as a continuum might be promising for generating new insights into the relationship with these levels.

Specifically, it has been mentioned above that a vehicle can also cooperate only on one or the other axis of control (e.g., on the longitudinal axis as a cooperative ACC), but to be truly cooperative, it should have the ability for lateral and longitudinal control, that is, because the automation needs an **understanding of the situation** and the humans actions in order to predict the outcomes of actions and provide support [21, 25]. Only on higher levels of automation, the automation also needs to be able to **execute lateral and longitudinal control** for cooperative guidance and control. Specifically, from Level 5 of automation of decision and action selection by Parasuraman et al. [51], SAE Level 2 [6] or partially automated according to BASt [29]. This claim needs to be fulfilled in order to enable cooperative driving, because the automation needs to execute these functions at least partially. Only for what is called autonomous driving here [10] the automation needs to be able to **execute lateral and longitudinal control at all times and in all situations** as it is no longer cooperating with a driver. Accordingly, different degrees of autonomy can be formulated and specifically the necessary degree of autonomy for different driving modes can be determined. Based on these considerations, it can be stated that:

In automated driving, autonomy of the automation can be seen as a prerequisite for a cooperative automation. However, different levels of automation presume different minimum degrees of autonomy. The minimum degree of autonomy for assisting the driver includes the ability to understand lateral and longitudinal control and factors, which influence these. For automated driving (i.e., from SAE Level 2 onwards), a degree of autonomy is needed which includes the ability to execute lateral and longitudinal control.

This is in line with the current nomenclature of automated and autonomous driving (c.f. [6, 29]) as these describe how much freedom to execute its autonomous capacities the automation is given. Moreover, with these insights the propositions on autonomy by Flemisch et al. [21] introduced above can be explained. They state that there is a tradeoff between cooperativeness and autonomy of a technical system, but also mentioned that some level of autonomy is necessary for cooperation. For cooperative driving at least a low degree of autonomy and for higher levels of automation a somewhat higher degree is necessary. Even if the automation possesses the highest abilities that would allow a high autonomy, this can be used in a cooperative way. Only if the automation is given the freedom to realize this autonomy (even if it is only on longitudinal or lateral control), then the driver can no longer cooperate with the vehicle concerning the driving task, but the vehicle is driving independently.

1.6 Design Recommendations and Avenues for Future Research

Several recommendations for designing the cooperation between driver and automation can be drawn from the theories and findings presented above. The research on teams informs the designer about the importance of autonomy for the driver's motivation. According to Hackman and Oldham [31] and Kirkman et al. [43], it seems to be beneficial for humans to feel autonomous in the way they execute their work. More research is needed to clarify whether this principle is also applicable to the driving task. It might even be the case that feeling autonomous in his decisions could prevent the effect of the driver scapegoating the automation and implicitly giving up responsibility [42, 50]. Maybe, to this end it is not necessary to change the degree of realized autonomy the driver possesses when driving cooperatively, but reminding her of the fact that she can always make her own decisions and is responsible. Especially because for current implementations of partially automated driving, legislation holds the driver responsible in all situations, and it is crucial to keep the driver aware of this fact. Moreover, when the development of automated and autonomous driving proceeds, the motivational aspect of autonomy could become relevant for a different goal, as raising motivation for driving might help preventing skill degradation.

In line with this, other characteristics and flaws of the human user should be kept in mind. For example, complacency bias has been described above, but other biases, heuristic reasoning, and individual differences between drivers as well as within drivers (e.g., emotions and moods) should be taken into account by system designers. As some of these factors are non-rational and dynamic, this can be a highly complex task, which needs to be accompanied by research.

Applied to real-world examples, the above described Tesla accident shows that the cooperation between the automation and the human is a key factor to avoid safety-critical situations in the driving task. Since both partners have a certain degree of realized autonomy, the misjudgment of the partner's capability could lead to significant dangers in automated driving. The driver in the Tesla crash probably overestimated the capabilities of the automation (also see Fig. 1.2). Thus even though he knew about the shortcomings of the autopilot, he did not observe the traffic and could not cover for the failures of the automation.

In the framework of levels of automation, realized autonomy of the partners can vary depending on the automation level. Indeed, the levels described by SAE and BASt [6, 29] demand different levels of realized autonomy from both cooperation partners. This is exemplary described in Fig. 1.3. Both partners should be aware of the system's configuration for sharing the driving task in different modes. A highly automated vehicle with a higher degree of realized autonomy, based on higher abilities, would need a lower degree of involvement of the driver and a higher degree of realized autonomy of the automation when driving in partially automated mode than when driving in highly automated mode. Accordingly, vehicles with various automation modes might need to clarify the required autonomy of the partner in a very explicit way on each automation mode.

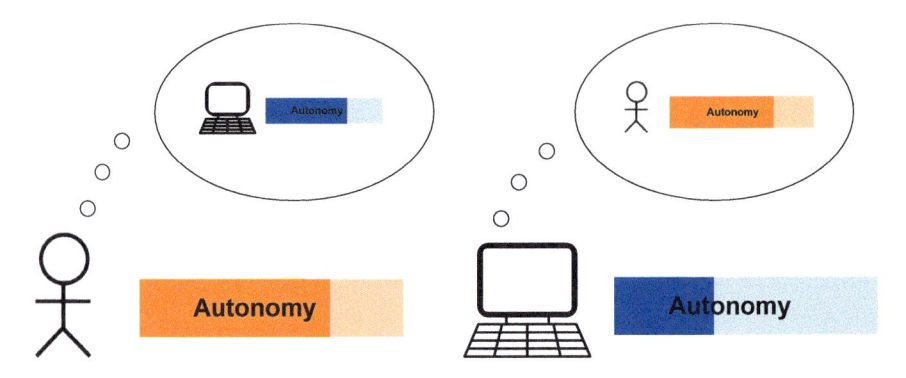

Fig. 1.2 Currently realized autonomy of human and automation and the estimate of realized autonomy by the other cooperation partner. Potential differences between the estimation and the actually realized autonomy due to misjudgment of the cooperation partner are shown, which could be highly critical and might lead to dangerous situations and crashes [24]

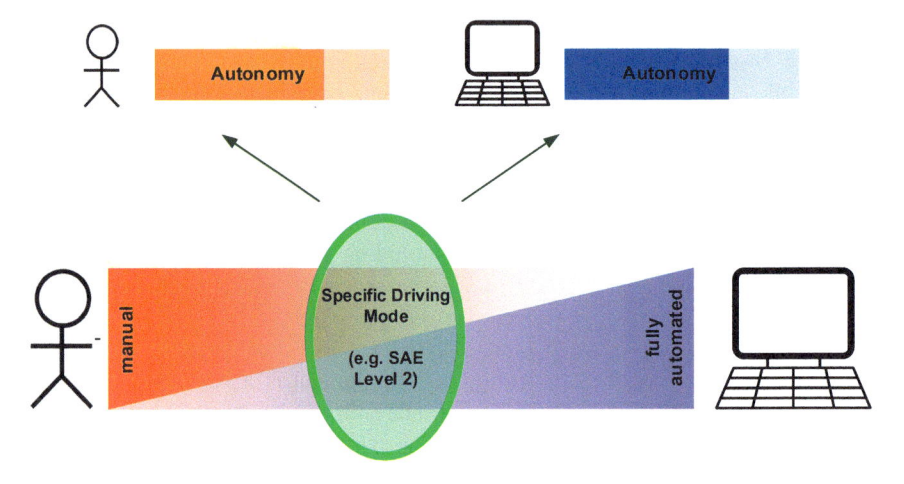

Fig. 1.3 Required realized autonomy for both partners at a specific level of automation. Elaborated on Flemisch et al. [20, 24, 26]

Other specific design recommendations can be derived from the conditions for cooperation presented by Axelrod [7]. His first condition (a likelihood of meeting in the future) is most complicated for transferring to the domain of cooperative driving. However, the automation could use hints that imply a future meeting. A basic implementation for this would be using phrases like "See you again" when being turned off. A learning automation on the other hand could ask for feedback after it has taken certain actions (or the driver has shown certain reactions). An example for this would be concerning greater safety margins for the driver's personal comfort. Moreover, Axelrod describes that a likelihood of meeting in the future promotes cooperation, because it increases accountability. Maybe drivers would be more willing to cooperate with the automation if they were more aware of their own accountability. The second condition (an ability to identify each other) points to the importance of making the automation's motivations and actions transparent and reliable and providing clear signs about the automation's willingness to cooperate. The last condition (a record of past behavior) emphasizes the importance of reputation and consistent functioning in order to promote cooperation of the driver. Only when previous experiences have been positive, enduring cooperation can be expected. As has been described above, this notion does not necessarily need to come from personal experience, but can also be based on reputation or recommendation by others.

Lastly, design recommendations can be derived from the working definition on cooperation presented above. The working definition of cooperation indicates that while the shared goal of reaching a destination safely is presumed to be universal, different sub-goals can be held by drivers and shared by the automation. If these individual goals could be communicated, the technical sub-system could optimally adapt to the driver's wishes, and therefore, cooperation could be expanded.

1.7 Implementation of the Design Recommendations and Working Definitions in the H-Mode

Considering the working definitions and some of the design recommendations elaborated above, a version of cooperative guidance and control [25] has been implemented in the H-Mode driving simulator. Based on the H-Metaphor [24], H-Mode focuses on the haptic-multimodal coupling between automation and driver. The H-Metaphor uses the example of rider or horse cart driver and horse as a blueprint for a cooperative relationship [23, 24]. Therefore, H-Mode is realized using haptically active interaction devices, when available. By providing bidirectional haptic interaction, a modality with a strong link to spatial movement can be addressed. Thus, the H-Mode driving simulator disposes of active actuators, such as active pedals and an active steering wheel, or active side sticks. These devices are used for providing a position or a force feedback, depending on the current driving mode and driving situation. This setup fulfills the recommendation derived from the second condition by Axelrod [7] to make the actions and intentions of the automation traceable by displaying the actions and intentions of the automation by haptic feedback.

The H-Mode consists of three different assistance and automation modes. In assisted mode (Tight Rein), the automation has direct coupling with the driver. While the human driver is still the leading partner in this cooperative setting, the automation provides support for longitudinal and lateral vehicle control. In Loose Rein, the automation stabilizes the vehicle, while the driver is still kept in the control loop by being permanently involved on the guidance level. Thus, she can decide for certain driving maneuvers and initiate those by using the active interfaces. For example, when driving in a highway scenario, the human driver could initiate a lane change maneuver by turning the steering wheel in the corresponding direction. If and when possible, the automation conducts this maneuver after the intention of the human driver is understood. In the highest H-Mode automation level (Secured Rein), the automation is temporarily fully in control. In this mode, the human driver can retreat herself from the dynamic driving task for a certain time. With these different automation modes, the driver can select how far the automation can realize its potential autonomy while having the chance to change this degree any time she wants. Additionally, the driver stays in the loop except in the case of Secured Rein; thus, the risk of complacency bias and scapegoating the automation is decreased [3, 25].

The transitions between the three described automation modes can be initiated by both, the driver or the automation. There are several possibilities for the driver to initiate those transitions. Firstly, grip-sensitive inceptors can be built in the device. Secondly, capacitive sensors or torque and force measurements can be used. Thirdly, there is a graphical interface with buttons, which can be used by the driver to initiate transitions between the modes. When relying on sensor technology, a transition to a higher automation mode is, for example, initiated, if the driver loosens her grip on the corresponding inceptor. Subsequently, by increasing the grip on the interface, the driver can initiate a transition to a lower automation mode. Thus, by grasping harder she gains more immediate control over the vehicle's guidance. These well-defined transition initiation conditions enable a high traceability of the automation's actions for the driver in the particular mode while the risk for mode confusion is limited.

The H-Mode automation is designed to increase the inner compatibility with the human goal and value system and to increase the compatibility with the way a human would drive [22], whereas the main focus is on the guidance and control levels. On the guidance layer, the automation recommends possible maneuvers on the haptic inceptor and additionally on the visual channel, whereby the driver chooses a recommended maneuver by conducting a steering gesture on the inceptor. The driver can also conduct maneuvers which are not recommended by the automation, as long as the maneuver is safe. The same condition applies also to the overruling regulation. The driver can override the automation if the planned action is safe. Thus, it is not possible for the driver to act against the common goal of the agents to reach the planned destination safely.

Both agents act through the inceptors, so that actions of both the partners become observable and traceable to the other partner. Since a minimum degree of cooperation between the driver and automation is required for the H-Mode, the resulting conflicts on the guidance and control level are solved by an arbitration mechanism.

The latest prototype of the H-Mode uses longitudinal and lateral automation with real-time maneuver and trajectory planning [3, 45], including lane departure warning, lane keeping and lane change assistance, full range ACC, forward collision warning and avoidance by braking or evasion functionality [25]. With these functionalities, the lowest degree of autonomy as specified above is enabled for the automation, which makes cooperative guidance and control possible. The HMI is either an active side stick or the combination of active steering wheel and pedals, which are equipped with a grip force sensing device. Additionally, a head-up display shows the current and alternative trajectories, whereas the automation modes are displayed on a separate touch screen (head-down display). Further information on cooperative guidance and control and on the H-Mode can be found in [3, 4, 22, 25].

1.8 Conclusion

Selected perspectives from different scientific fields have been presented in order to develop a working definition of cooperation and clarify the role of autonomy in the domain of automated driving. Moreover, the relationship between the two concepts has been specified. Design recommendations for driver–vehicle cooperation have been presented and in part related to the Tesla vehicle crash. Finally, the H-Mode prototype has been described as an implementation example of the definition and design recommendations.

Most importantly, it has been demonstrated that an automation can be seen as a cooperation partner and that driver–automation cooperation entails the common goal of reaching a destination safely and building and executing plans that are agreed on by both agents. Autonomy of the automation can be seen as a meta-ability and can be divided into potential and realized autonomy. This meta-ability to act and control a situation without the support of other agents is based on sufficient abilities to act. Different assistant and automation modes define the authority and, combined with sufficient ability, the autonomy of each partner. Different levels of automation presume a different minimum degree of realized autonomy. Design recommendations concerning cooperative driving mainly focus on making the automation as transparent and reliable as possible and signaling cooperativeness. Moreover, it should be adaptive to inter- and intra-individual differences and the designer should mind general human characteristics and biases.

Concerning the definition, findings, and recommendations, it must be noted that the presented perspectives, theories, and research findings from different areas are not based on an exhaustive literature review. Indeed, different definitions of the concepts of autonomy and cooperation might be found, which could further inform the work in this area.

Nevertheless, the presented definition, findings, and design recommendations elaborate on present work and provide more information and clarity on the central concepts in the domain of automated driving. Furthermore, the presented insights can be applied to different areas of human–machine cooperation, for example, to cooperative automation in flying, robotics, or production.

References

1. AAIC: REPORT on the investigation into the crash of A310-308, registration F-OGQS, on 22 March 1994 near the city of Mezhdurechensk. aviation-safety.net (1994)
2. Ackerman, E.: Fatal Tesla self-driving car crash reminds us that robots aren't perfect [http://spectrum.ieee.org/cars-that-think/transportation/self-driving/fatal-tesla-autopilot-crash-reminds-us-that-robots-arent-perfect]. Website (2016). URL http://spectrum.ieee.org/cars-that-think/transportation/self-driving/fatal-tesla-autopilot-crash-reminds-us-that-robots-arent-perfect
3. Altendorf, E., Baltzer, M., Heesen, M., Kienle, M., Weißgerber, T., Flemisch, F.: H-mode. In: H. Winner, S. Hakuli, F. Lotz, C. Singer (eds.) Handbook of Driver Assistance Systems:

Basic Information, Components and Systems for Active Safety and Comfort, pp. 1499–1518. Springer, Cham (2016). https://doi.org/10.1007/978-3-319-12352-3_60
4. Altendorf, E., Weßel, G., Baltzer, M., Canpolat, Y., Flemisch, F.: Joint decision making and cooperative driver-vehicle interaction during critical driving situations. i-com 15(3), 265–281 (2016)
5. Altendorf, E., Wessel, G., Baltzer, M., Canpolat, Y., Flemisch, F.: Joint decision making and cooperative driver-vehicle interaction during critical driving situations. i-com J. Interact Media (2017)
6. road Automated Vehicle Standards Committee, O., et al.: Sae j3016: Taxonomy and definitions for terms related to on-road motor vehicle automated driving systems. SAE International
7. Axelrod, R.M.: The evolution of cooperation. Basic Books (1984)
8. Babiceanu, R.F., Chen, F.: Development and applications of holonic manufacturing systems: a survey. J Intell. Manuf. 17, 111–131 (2006)
9. Bahner, J.: Übersteigertes Vertrauen in Automation: Der Einfluss von Fehlererfahrungen auf Complacency und Automation Bias. Ph.D. thesis, Technische Universitt Berlin, Fakultt V - Verkehrs- und Maschinensysteme (2008)
10. Barringer, K.L.: Code Bound and down-a long way to go and a short time to get there: autonomous vehicle legislation in Illinois. S. Ill. ULJ **38**, 121 (2013)
11. BEA: Final Report On the accident on 1st June 2009 to the Airbus A330-203 registered F-GZCP operated by Air France flight AF 447 Rio de Janeiro - Paris; Bureau dEnqutes et dAnalyses. www.skybrary.aero (2012)
12. Boca, R., Boca, M., Dlhan, L., Falk, K., Fuess, H., Haase, W., Jaroščiak, R., Papankova, B., Renz, F., Vrbova, M., et al.: Strong cooperativeness in the mononuclear iron (ii) derivative exhibiting an abrupt spin transition above 400 k. Inorg. Chem. **40**(13), 3025–3033 (2001)
13. Butler, R.W., Miller, S.P., Potts, J.N., Carreno, V.A.: A formal methods approach to the analysis of mode confusion. In: Digital Avionics Systems Conference, 1998. Proceedings., 17th DASC. The AIAA/IEEE/SAE, vol. 1, pp. C41–1. IEEE (1998)
14. Christensen, J.H.: Holonic manufacturing systems: initial architecture and standards directions. Proceedings of 1st Euro Workshop on Holonic Manufacturing Systems (1994)
15. Christman, J.: Relational autonomy, liberal individualism, and the social constitution of selves. Philos. Stud. **117**(1), 143–164 (2004)
16. Dietrich, E.: Thinking computers and virtual persons: essays on the intentionality of machines. Academic Press, Cambridge (2014)
17. Endsley, M.R.: Designing for situation awareness: an approach to user-centered design. CRC Press, Boca Raton (2016)
18. Fiehler, R.: Kommunikation und Kooperation. Theoretische und empirische Untersuchungen zur kommunikativen Organisation kooperativer Prozesse. Linguistik Band 4. Einhorn-Verlag, Berlin (1980)
19. Flemisch, F., Abbink, D., Itoh, M., Pacaux-Lemoine, M.P., Weßel, G.: Shared control is the sharp end of cooperation: Towards a common framework of joint action, shared control and human machine cooperation. IFAC-PapersOnLine **49**(19), 72–77 (2016)
20. Flemisch, F., Altendorf, E., Canpolat, Y., Weßel, G., Baltzer, M., López, D., Herzberger, N., Voß, G., Schwalm, M., Schutte, P.: Uncanny and unsafe valley of assistance and automation: First sketch and application to vehicle automation. In: Schlick C.M., Duckwitz S., Flemisch F., Frenz M., Kuz S., Mertens A., Mütze-Niewöhner S. (eds.) Advances in Ergonomic Design of Systems, Products and Processes., Proceedings of the Annual Meeting of GfA. Springer, Berlin (2016)
21. Flemisch, F., Baltzer, M., Altendorf, E., López, D., Rudolph, C.: Kooperativität und Arbitrierung versus Autonomie: Grundsätzliche Überlegungen zur kooperativen Automation mit anschaulichen Beispielen. 57. Fachausschusssitzung Anthropotechnik der DGLR: Kooperation und kooperative Systeme in der Fahrzeug-und Prozessfhrung (2016)
22. Flemisch, F., Heesen, M., Hesse, T., Kelsch, J., Schieben, A., Beller, J.: Towards a dynamic balance between humans and automation: authority, ability, responsibility and control in shared and cooperative control situations. Cogn. Technol. Work **14**(1), 3–18 (2012). https://doi.org/10.1007/s10111-011-0191-6

23. Flemisch, F., Heesen, M., Kelsch, J., Schindler, J., Preusche, C., Dittrich, J.: Shared and cooperative movement control of intelligent technical systems: Sketch of the design space of haptic-multimodal coupling between operator, co-automation, base system and environment. IFAC Proc. Volumes **43**(13), 304–309 (2010)
24. Flemisch, F.O., Adams, C.A., Conway, S.R., Goodrich, K.H., Palmer, M.T., Schutte, M.C.: The H-Metaphor as a Guideline for Vehicle Automation and Interaction. Report No. NASA/TM-2003-212672. Tech. rep., Hampton, NASA Research Center (2003)
25. Flemisch, F.O., Bengler, K., Bubb, H., Winner, H., Bruder, R.: Towards a cooperative guidance and control of highly automated vehicles: H-mode and conduct-by-wire. Ergonomics **57**(3), 343–360 (2014)
26. Flemisch, F.O., Kelsch, J., Löper, C., Schieben, A., Schindler, J.: Automation spectrum, inner/outer compatibility and other potentially useful human factors concepts for assistance and automation. In: de Waard D., Flemisch F.O., Lorenz B., Oberheid H., Brookhuis K.A. (eds.) (2008), Human Factors for assistance and automation, pp. 1–16. Maastricht, the Netherlands: Shaker Publishing (2008)
27. Friedman, B.: Moral responsibility and computer technology. (1990)
28. Friedman, B.: It's the computer's fault?: reasoning about computers as moral agents. In: Conference companion on Human factors in computing systems, pp. 226–227. ACM (1995)
29. Gasser, T.M., Arzt, C., Ayoubi, M., Bartels, A., Brkle, L., Eier, J., Flemisch, F., H"acker, D., Hesse, T., Huber, W., Lotz, C., Maurer, M., Ruth-Schumacher, S., Schwarz, J., Vogt, W.: Rechtsfolgen zunehmender Fahrzeugautomatisierung - Gemeinsamer Schlussbericht der Projektgruppe. Fahrzeugtechnik F 83, Bundesanstalt für Straßenwesen (BASt) (2012)
30. Haberfellner, R., de Weck, O., Fricke, E., Vössner, S.: Systems Engineering-Principles and Application. Orell Füssli, Systems Engineering-Grundlagen und Anwendung. Zürich (2012)
31. Hackman, J.R., Oldham, G.R.: Motivation through the design of work: test of a theory. Organizational behavior and human performance **16**(2), 250–279 (1976)
32. Hamilton, W.D., Axelrod, R.: The evolution of cooperation. Science **211**(27), 1390–1396 (1981)
33. Hasegawa, T., Seki, T., Tamura, S.: Environment for developing Holonic manufacturing system applications-HMS shell. Intell. Auton. Syst. **6**, 379 (2000)
34. Heesen, M., Kelsch, J., Lper, C., Flemisch, F.: Haptisch-multimodale Interaktion fr hochautomatisierte, kooperative Fahrzeugfhrung bei Fahrstreifenwechsel-, Brems- und Ausweichmanvern. In: 11. Braunschweiger Symposium Automatisierungs-, Assistenzsysteme und eingebettete Systeme fr Transportmittel (AAET) (2010)
35. Hoc, J.M.: From human - machine interaction to human - machine cooperation. Ergonomics **43**(7), 833–843 (2000)
36. Hoc, J.M., Mars, F., Milleville-Pennel, I., Jolly, É., Netto, M., Blosseville, J.M.: Human-machine cooperation in car driving for lateral safety: delegation and mutual control. Le travail humain **69**(2), 153–182 (2006)
37. Hollnagel, E., Woods, D.D.: Joint cognitive systems: foundations of cognitive systems engineering. CRC Press, Boca Raton (2005)
38. Holzmann, F.: Adaptive cooperation between driver and assistant system. In: Adaptive Cooperation between Driver and Assistant System, pp. 11–19. Springer, Berlin (2008)
39. Hughes, D., Dornheim, M.: Automated cockpits: whos in charge. Aviation Week & Space Technology (1995)
40. Johnson, D.G.: Computer systems: moral entities but not moral agents. Ethics Inf. Technol. **8**(4), 195–204 (2006)
41. Kaber, D.B.: All that automation is... and all that autonomy should not be. Presentation given at TNO Symposium on Human Factors in Automated Systems, Soesterberg, The Netherlands (2016)
42. Kim, T., Hinds, P.: Who should I blame? Effects of autonomy and transparency on attributions in human-robot interaction. In: ROMAN 2006-The 15th IEEE International Symposium on Robot and Human Interactive Communication, pp. 80–85. IEEE (2006)

43. Kirkman, B.L., Rosen, B., Tesluk, P.E., Gibson, C.B.: The impact of team empowerment on virtual team performance: the moderating role of face-to-face interaction. Acad. Manage. J. **47**(2), 175–192 (2004)
44. Koestler, A.: The Act of Creation. Penguin Books, New York (1964)
45. Löper, C., Flemisch, F.O.: Ein Baustein für hochautomatisiertes Fahren: Kooperative, manöverbasierte Automation in den Projekten H-Mode und HAVEit. In: 6. Workshop Fahrerassistenzsysteme, pp. 136–146. Freundeskreis Mess-und Regelungstechnik Karlsruhe eV (2009)
46. NHTSA: Preliminary statement of policy concerning automated vehicles. Washington, DC pp. 1–14 (2013)
47. NTSB: National Transportation Safety Board: Aviation Accident Report NTSB-AAR-73-14. Tech. rep., NTSB (1973)
48. Oxford-Dictionaries: (2017). https://en.oxforddictionaries.com/definition/cooperation
49. Pacaux-Lemoine, M.P., Debernard, S.: Common work space for human - machine cooperation in air traffic control. Control Eng. Pract. **10**(5), 571–576 (2002)
50. Parasuraman, R., Molloy, R., Singh, I.L.: Performance consequences of automation-induced'complacency. Int. J. Aviat. Psychol. **3**(1), 1–23 (1993)
51. Parasuraman, R., Sheridan, T.B., Wickens, C.D.: A model for types and levels of human interaction with automation. IEEE Trans. Syst. Man Cybern. **30**(3), 286–297 (2000)
52. Pollard, E., Morignot, P., Nashashibi, F.: An ontology-based model to determine the automation level of an automated vehicle for co-driving. In: Information Fusion (FUSION), 2013 16th International Conference on, pp. 596–603. IEEE (2013)
53. Probst, G.J., Ulrich, H.: Anleitung zum ganzheitlichen Denken und Handeln. Haupt, Bern (1988)
54. Rammert, W.: Technik in Aktion: verteiltes Handeln in soziotechnischen Konstellationen. (TUTS- Working Papers 2-2003) (2003)
55. Rammert, W.: Technik als verteilte Aktion: Wie technisches Wirken als Agentur in hybriden Aktionszusammenhängen gedeutet werden kann. TechnikHandelnWissen: Zu einer pragmatistischen Technik-und Sozialtheorie pp. 79–89 (2007)
56. Rammert, W.: Technik und Handeln: Wenn soziales Handeln sich auf menschliches Verhalten und technische Abläufe verteilt. TechnikHandelnWissen: Zu einer pragmatistischen Technik-und Sozialtheorie pp. 91–123 (2007)
57. Rammert, W., Schulz-Schaeffer, I.: Technik und Handeln-Wenn soziales Handeln sich auf menschliches Verhalten und technische Artefakte verteilt. (TUTS - Working Papers 4-2002) (2002)
58. Rasmussen, J.: Skills, rules and knowledge; signals, signs, and symbols, and other distinctions in human performance models. IEEE Trans. Syst. Man Cybern. **3**(3), 257–266 (1983)
59. Rödel, C., Stadler, S., Meschtscherjakov, A., Tscheligi, M.: Towards autonomous cars: the effect of autonomy levels on acceptance and user experience. In: Proceedings of the 6th International Conference on Automotive User Interfaces and Interactive Vehicular Applications, pp. 1–8. ACM (2014)
60. Salas, E., Burke, C.S., Bowers, C.A., Wilson, K.A.: Team training in the skies: does crew resource management (crm) training work? Hum. Factors: J. Hum. Factors Ergon. Soc. **43**(4), 641–674 (2001)
61. Sanyal, B.S.: Cooperative autonomy: The dialectic of State-NGOs relationship in developing countries. International Institute for Labour Studies, (1994)
62. Sarter, N.B., Woods, D.D.: How in the world did we ever get into that mode? mode error and awareness in supervisory control. Hum. Factors **37**(1), 5–19 (1995)
63. Savage, I.: Comparing the fatality risks in United States transportation across modes and over time. Res. Transp. Econ. **43**(1), 9–22 (2013)
64. Schilcher, C., Schmiede, R., Will-Zocholl, M., Ziegler, M.: Vertrauen und Kooperationen in einer sich wandelnden Arbeitswelt–eine Einführung. In: Vertrauen und Kooperation in der Arbeitswelt, pp. 11–19. Springer, Berlin (2012)
65. Shladover, S.E.: Cooperative (rather than autonomous) vehicle-highway automation systems. IEEE Intell. Transp. Syst. Mag. **1**(1), 10–19 (2009)

66. Snyder, C.R., Lassegard, M., Ford, C.E.: Distancing after group success and failure: basking in reflected glory and cutting off reflected failure. J. Pers. Soc. Psychol. **51**(2), 382 (1986)
67. Spinda, J.S.: The development of Basking in Reflected Glory BIRGing) and Cutting off Reflected Failure (CORFing) measures. J. Sport Behav. **34**(4), 392 (2011)
68. United Nations, E., Council, S.: Report of the sixty-eighth session of the working party on road traffic safety. In: Economic Commission for Europe, Inland Transport Committee, Working Party on Road Traffic Safety (2014)
69. Vanderhaegen, F.: Autonomy control of human-machine systems. IFAC Proc. Volumes **43**(13), 398–403 (2010)
70. Wann, D.L., Hamlet, M.A., Wilson, T.M., Hodges, J.A.: Basking in reflected glory, cutting off reflected failure, and cutting off future failure: the importance of group identification. Soc. Behav. Pers.: Int. J. **23**(4), 377–388 (1995)
71. Wiener, E.: Complacency: Is the term useful for air safety. In: Proceedings of the 26th Corporate Aviation Safety Seminar, vol. 117 (1981)
72. Wiener, E.L., Kanki, B.G., Helmreich, R.L.: Crew resource management. Academic Press, Cambridge (2010)

Chapter 2
Robust Real-World Emissions by Integrated ADF and Powertrain Control Development

Frank Willems, Peter van Gompel, Xander Seykens and Steven Wilkins

Abstract This work gives an outlook on the potential of automated driving functions (ADFs) to reduce real-world CO_2 and pollutant emissions for heavy-duty powertrains. Up to now, ADF research mainly focuses on increased traffic safety, driver comfort, and road capacity. Studies on emissions are lacking. By taking the driver out-of-the-loop, cycle-to-cycle variability is removed and energy losses and large accelerations can be significantly reduced. This enhances emission performance robustness, which will allow for more fuel-efficient engine settings. A general, optimal control framework is introduced, which integrates ADF with energy and emission management. Based on predictions of the vehicle power demand and emissions, a desired vehicle velocity profile, which minimizes the overall vehicle energy consumption, is determined. In this approach, real-world tailpipe emissions are explicitly taken into account. This opens the route to emission trading on vehicle or even, on platoon, fleet, and traffic level. For the combined ADF and powertrain development, testing, and certification, various opportunities are presented to fully exploit the synergy between these systems and to reduce development time and costs. By equipping vehicles with an emission monitoring system, real-world data of the ADF emission reduction potential becomes available. As validated traffic and component aging models are lacking, this data is also valuable for realistic scenario development and uncertainty modeling in virtual or mixed testing. This will lead to improved robustness evaluation and performance.

F. Willems (✉) · P. van Gompel · X. Seykens · S. Wilkins
TNO Automotive, Powertrains, Helmond, The Netherlands
e-mail: Frank.Willems@tno.nl

P. van Gompel
e-mail: Peter.van.Gompel@tno.nl

X. Seykens
e-mail: Xander.Seykens@tno.nl

S. Wilkins
e-mail: Steven.Wilkins@tno.nl

F. Willems
Faculty of Mechanical Engineering, Eindhoven University of Technology, Eindhoven, The Netherlands

2.1 Introduction

Development of advanced driver assistance systems (ADASs) and automated driving functions (ADFs) mainly focuses on reducing the number of (fatal) accidents and on increasing the traffic flow throughput. Also, increased comfort by reduced driver's workload is examined. As illustrated in Fig. 2.1, these systems also have potential to minimize energy consumption (and thus CO_2) and to reduce pollutant emissions. In the field of ADAS, speed and gear ratio selection advise and tire pressure monitoring and control systems are examples of systems that focus on fuel consumption reduction. Also, navigation systems with time or fuel-optimal route advise and ecomodes for gear shifting are available. ADF studies, with automatically controlled vehicle speed, mainly focus on adaptive cruise control (ACC) and cooperative ACC (CACC) applications, including platooning [2, 14]. In general, these ADF studies pay minimal attention to the impact on fuel consumption (and thus CO_2) and, especially, on pollutant emissions.

This study discusses the reduction potential of real-world CO_2 and pollutant emissions in case the driver is taken out-of-the-loop (SAE automation level 2 and higher). We concentrate on on-road, heavy-duty (HD) powertrains, where platooning is foreseen to be introduced around 2020 for long-haul type of applications in green corridors. Due to their unmatched energy density, efficiency, and durability, diesel engines are expected to stay the main propulsion source for the upcoming decades for long-haul trucks. However, limits on real-world emissions will become increasingly strict. Note that the introduced ideas are also applicable to light-duty and off-road, heavy-duty applications. To meet the recently introduced real-world driving emission (RDE) limits, light-duty applications can benefit from the experience and knowledge that is already gained in the heavy-duty field.

This work is organized as follows. First, the main drivers and challenges for future automotive powertrain development are reviewed. Section 2.3 discusses the ADF

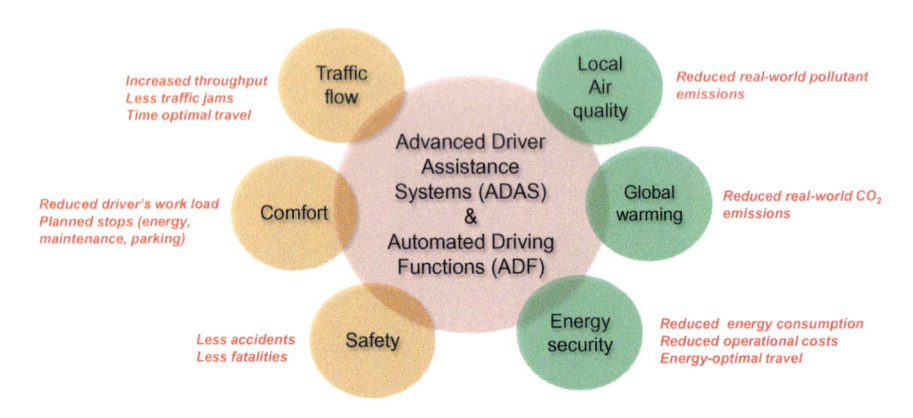

Fig. 2.1 Overview of potential added value of ADAS and ADF

potential for reduction of real-world driving fuel consumption, CO_2, and pollutant emissions. A general optimal control framework is introduced that extends existing energy and emission management strategies for ADF. Section 2.4 presents opportunities for synergy in joint ADF and powertrain development and testing. Finally, the main findings are summarized in Sect. 2.5.

2.2 Drivers and Challenges in HD Powertrain Development

2.2.1 Real-World Pollutant Emissions

Over the last two decades, powertrain development is mainly driven by legislation. Due to growing societal concern about air quality, increasingly strict targets are introduced. Focus was initially on reducing pollutant emissions, i.e., nitrogen oxides (NO_x), particulate matter (PM), hydrocarbons (HC), and carbon monoxide (CO) emissions, in test cycles under nominal, well-defined conditions in the laboratory. Compared to the Euro-I levels introduced in 1992, this resulted in 95% and 97% reduction in NO_x and PM emissions, respectively, for the current Euro-VI levels. As similar reduction levels were not observed on the road, attention shifted more recently to meet real-world emission targets and to monitor on-road performance. With Euro-VI, additional requirements have to be met to demonstrate in-service conformity (ISC) as well as to perform more extensive onboard diagnostics (OBD). These measures appeared to be very effective and will be extended to a wider range of operating conditions in the future.

Real-world driving emissions can vary significantly due to, e.g., varying duty cycles (application, weight), varying ambient conditions, aging and wear of components, driver behavior, and traffic flow. Figure 2.2a shows the effect of the duty cycle on emission performance. The applied Integrated Emission Management (IEM)

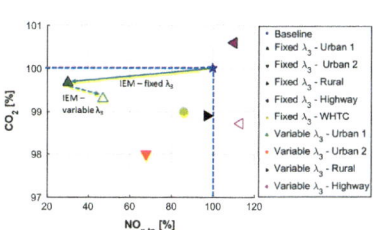

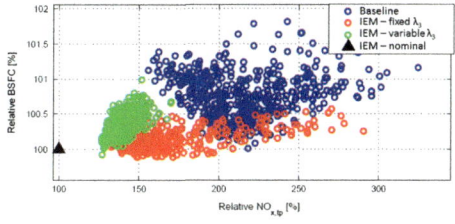

(a) Impact of duty cycle (drawn from data found in [16]). Symbols: solid - fixed control setting; open - variable control setting

(b) Impact of sensor and actuator uncertainty for real-world urban cycle. Nominal case without uncertainty [19].

Fig. 2.2 Illustration of real-world variations in total fluid costs (TFC) and normalized tailpipe NO_x emissions (baseline = Euro-VI production strategy; IEM = Integrated Emission Management strategy)

strategy minimizes the operational costs for the combined engine-after treatment system. It was calibrated for the type-approval cycle: so-called World Harmonized Test Cycle (WHTC). This controller gives a large variation in NO_x–CO_2 performance for real-world urban, rural, and highway cycles (fixed cases). Figure 2.2b illustrates the impact of sensor and actuator uncertainty on total fluid costs (TFC), which are related to fuel and AdBlue consumption, and on tailpipe NO_x conformity factor for a real-world urban cycle. Similar to the previous case, fixed controller settings results in wide variation of NO_x emissions. These examples illustrate the need for robust powertrain performance.

In case route information is available, the performance could be further optimized by adaptation of control parameters (i.e., λ_3 in the applied IEM strategy) with respect to upcoming drive pattern. The corresponding results are indicated by the variable λ_3 cases. This demonstrate the potential of an adaptive approach: improved TFC-NO_x trade-off and reduced NO_x variations, see Fig. 2.2a, b, respectively.

2.2.2 Fuel Consumption and CO_2 Emissions

More recently, increasing efforts to minimize global warming has led to regulation of greenhouse gas (GHG) emissions for automotive applications. Up to now, focus is on CO_2 emissions, but future inclusion of methane (CH_4) and nitrous oxide (N_2O) is under consideration. These components have a global warming potential (GWP) that is approx. 25 and 298 larger compared to CO_2, respectively. For on-road HD applications, with GHG Phase 2 the USA is front-runner in the implementation of CO_2 related legislation: 40% fuel consumption (FC) reduction in 2027 compared to 2010 [18]. Besides these targets that hold on vehicle level, engines are additionally required to reduce CO_2 emissions by 5% compared to GHG Phase 1 levels. In Europe, proposed targets (20% CO_2 reduction in 2020 by ACEA) and testing methodology are still under debate.

Automotive industry faces enormous challenges to simultaneously meet pollutant emission targets and reduces fuel consumption (and CO_2 emissions); since the introduction of increasingly strict emission levels in 1992, manufacturers struggled to find a cost-efficient solution without increasing fuel consumption. As shown in [7], over the last two decades, real-world fuel consumption stabilized around 35 [l/100km] for the studied trucks over the specified real-world driving cycle.

2.2.3 Powertrain Complexity

To meet the targets set by legislation, automotive powertrains have become increasingly complex. Engine measures, such as common rail fuel injection equipment with multi-pulse fueling capability, variable turbine geometry (VTG), and cooled exhaust gas recirculation (EGR) systems, are introduced. This is combined with exhaust gas aftertreatment systems with diesel particulate filters (DPFs) and urea-based selective catalytic reduction (SCR) systems. In order to improve the energy efficiency, energy recovery technologies, such as hybrid electric drivetrains and waste heat recovery (WHR) systems, are studied. These trends set demanding requirements for the development and testing process; due to the increased number of subsystems and actuators, it is no longer straightforward to maximize powertrain performance under varying operating conditions and maintain development time and costs at acceptable levels.

2.3 ADF Potential for Robust Powertrain Performance

Besides powertrain-related measures, such as engine efficiency and carbon fuel content, vehicle measures and improved transport efficiency are important focus areas to reduce tank-to-wheel CO_2 emissions for truck applications:

$$\underbrace{CO_2}_{\text{(g/ton km)}} = M_{CO2,fuel} \cdot \underbrace{\frac{1}{\eta_{engine}} \cdot \frac{1}{\eta_{drivetrain}}}_{\text{(g/kWh)}} \cdot \underbrace{W_{vehicle}}_{\text{(kWh)}} \cdot \underbrace{\frac{1}{\eta_{transport}}}_{\text{(1/ton km)}} \quad (2.1)$$

where $M_{CO2,fuel}$ is the fuel depend, specific CO_2 emission and $W_{vehicle}$ is the combination of road load and required power to drive auxiliaries over the studied trip:

$$W_{vehicle} = \int_0^{t_e} \left(F_{acc}(m, \dot{v}) + F_{air}(v, d_{veh}) + F_{roll}(m, \alpha) + F_{gravity}(m, \alpha) + F_{aux} \right) v dt$$

with gross mass m, road slope α, distance d_{veh} to the front vehicle, vehicle speed v, and vehicle acceleration $\dot{v}$.

By taking the driver out-of-the-loop, ADF creates opportunities to minimize energy consumption due to increased flexibility in the selection of v, d_{veh}, and F_{aux} (assuming supply on demand) for a given route and mass. By automation, a fuel-optimal vehicle velocity profile can be realized. For ACC applied to passenger cars, up to 16% fuel reduction is claimed in [14] for a 20 [s] prediction window and $d_{veh,min}$=20 [m]. Besides city and urban driving applications, there is also potential for long-haul applications. Figure 2.3b, c shows two highway examples: national (part of TEN-T toward Rotterdam Harbor (highway driving—flat)) and Trans-European highway driving (highway driving—hilly), respectively. In both highway cases, significant speed variations are observed. Fuel savings up to 6.2% are reported for a long-haul truck in cruise control conditions on a hilly highway [11]. Reducing acceleration events also minimizes peaks in pollutant emissions: so-called peak shaving. This, together with reduced cycle-to-cycle variations, will enhance emission performance robustness. Consequently, safety margins can be reduced, which allow for more fuel-efficient control settings. In [2], the effect of d_{veh} for truck platooning is studied: a maximum of 4.7–7.7% FC reduction depending on the time gap is reported.

(a) Urban driving - flat (b) Highway driving - flat (c) Highway driving - hilly

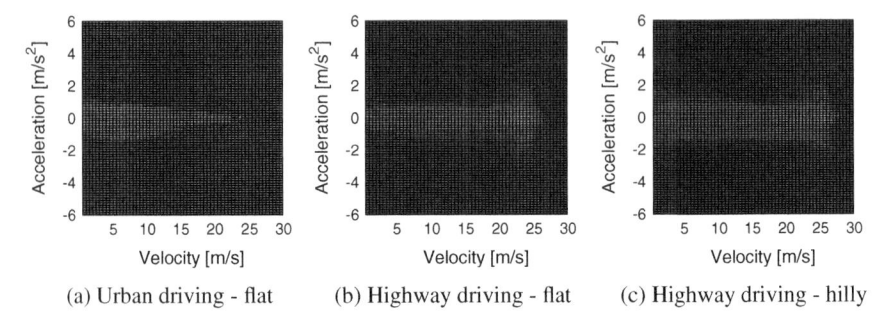

Fig. 2.3 Illustration of real-world speed variations. Graphs show recorded velocity–acceleration profiles of trucks for different duty cycles. Colors indicate frequency occurrence: ranging from blue (lowest frequency) to yellow (highest frequency)

For systems with buffers, energy and emission trading can be applied. So far, this trading is mainly limited to energy management on vehicle level using route information. In case of hybrid electric vehicles (HEV), up to 16% FC reduction is realized for simulated hilly FTP-75 conditions with a distribution truck on a chassis dynamometer [12]. Emissions are typically not included in these studies. In [26], a first attempt is made to explicitly deal with tailpipe emissions. Willems et al. achieved 2% FC reduction by online EGR–SCR balancing without prediction over a hot WHTC for a Euro-VI HD engine. Adaptive strategies based on route information will enable further performance optimization, as illustrated in Fig. 2.2. By moving the system optimization level from vehicle toward platoon, fleet, or traffic level, trading of energy and emissions become feasible. In [17], Murgovski et al. predicted 10% FC benefit for cooperative energy management applied to a platoon of four trucks running at constant cruising speed.

The potential ADF benefits depend on various factors. For the semi-automation case of platooning, the vehicle speed can be optimized and cycle-to-cycle variability due to the driver is removed. Although an increasing amount of static and dynamic route information becomes real-time available, including hills, traffic flow, and traffic light status, uncertainty in the future power request will still be significant. More precisely, actual traffic flow can still give large variations, since the level of ADF penetration also plays an important role; for many years, mixed traffic conditions (non-autonomous / autonomous) will occur.

Currently, fuel consumption and driver wages equally contribute to 60% of the total operational costs for a long-haul truck [1]. However, the location in the platoon affects the fuel saving potential and the driver's role. New business or operational models that share (financial) benefits over the platoon participants will stimulate the introduction of these technologies. Obviously, the full automation case will change the optimization problem significantly; as the driver's role (and the associated costs) is supposed to be minimized, this will give more freedom in travel time and speed range. By minimizing uncertainties in the predicted $\hat{W}_{veh}(t)$, the potential of online

2.3.1 Integration of ADF in Powertrain Control Strategy

Figure 2.4 shows the proposed integration of ADF with the powertrain control system. It is an extension of an existing integrated emission and energy management strategy [16]. For high automation levels, the driver's input is limited to the desired destination and travel time. In order to deal with the high number of inputs, a bi-optimization approach is proposed. The vehicle energy management determines the optimal vehicle velocity profile $v_{des}(t)$ that minimizes vehicle operational costs within safety and emission limits for the planned route. This requires information on current and future traffic flow, target (platoon or fleet) tailpipe emissions, and energy consumption. Based on $v_{des}(t)$ and the specified gear ratio $i_{tr,des}$, the corresponding, desired driving rotational speed $\omega_{d,des}$ and torque $\tau_{d,des}$ are given. Together with the

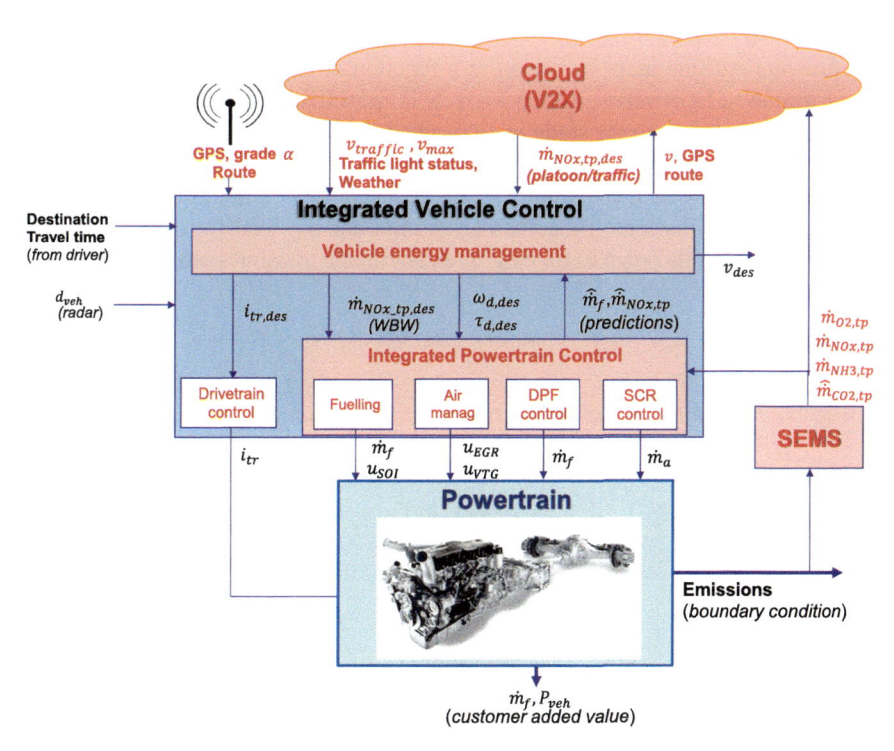

Fig. 2.4 Scheme of the proposed powertrain control system for automated driving. This illustrates the interaction between the powertrain controller, vehicle controller, and the cloud (adapted from [24])

tailpipe emission target, these are the inputs to the integrated powertrain controller, which realizes optimal engine and aftertreatment settings.

2.3.2 Optimal Control Problem

Similar to earlier work [16, 26], this control problem is formulated within the framework of optimal control. First, the cost function J (in Euro/s) is defined, which combines the labor costs c_{driver} and operational costs related to fuel mass flow $\dot{m}_f$ and AdBlue mass flow $\dot{m}_a$:

$$J(t) = c_{driver} + c_f \cdot \dot{m}_f(\omega_e, \tau_e, u_{EGR}, u_{VTG}, u_{SOI}) \\ + c_a \cdot \dot{m}_a(\omega_e, \tau_e, u_{EGR}, u_{VTG}, u_{SOI}) \quad (2.2)$$

where the AdBlue consumption is assumed to depend on SCR conversion efficiency η_{SCR} and engine out NO$_x$ emission $\dot{m}_{NOx,e}$, according to:

$$\dot{m}_a = 2.0067 \cdot \eta_{SCR} \cdot \dot{m}_{NOx,e}(\omega_e, \tau_e, u_{EGR}, u_{VTG}, u_{SOI}) \quad (2.3)$$

The fuel and AdBlue maps are given as a function of engine rotational speed ω_e, engine torque τ_e, EGR valve position u_{EGR}, VTG position u_{VTG}, and diesel injection timing u_{SOI}. Contrary to the study in [11], which minimizes travel time and energy consumption, we use actual operational costs, which explicitly includes the effect of aftertreatment. Note that minimizing fuel consumption is in close relation with minimizing CO$_2$ emissions. Then, for a given route and pay load, the optimal vehicle control problem boils down to (in case of an automatic transmission):

$$\min_{v,\, i_{tr}} \int_0^{t_e} J(t)dt \quad (2.4)$$

with decision variables: vehicle velocity v and transmission gear ratio i_{tr}, and subject to:

$$\dot{x} = f(x, u, t) \quad (2.5)$$

$$d_{veh} > d_0 + v \cdot h \quad (2.6)$$

$$v_{min} \leq v \leq v_{max} \quad (2.7)$$

$$a_{min} \leq \frac{dv}{dt} \leq a_{max} \quad (2.8)$$

$$\Phi(NO_{x\_tp,WBW} < CF90 \cdot Z_{NOx,TA}) \geq 0.9 \quad \text{(ISC limit)} \tag{2.9}$$

where:

$$NO_{x\_tp,WBW} = \frac{\int_0^{t_{WBW}} \dot{m}_{NOx\_tp}\, dt}{\int_0^{t_{WBW}} P_e\, dt}$$

The system dynamics, Eq.(2.5), are given by (see also [11, 20]):

$$\frac{dv}{dt} = \frac{1}{m_t}\left[\frac{i_{tr}\eta_{tr}}{r_w}\tau_d - F_{brake} - F_{aux} - \frac{c_D(d_{veh})}{2}A_a\rho_a v^2 - mg(c_r\cos\alpha + \sin\alpha)\right] \tag{2.10a}$$

$$\frac{dT_{DOC}}{dt} = c_{exh}\cdot \dot{m}_{exh}[T_{exh} - T_{DOC}] \tag{2.10b}$$

$$\frac{dT_{SCR}}{dt} = c_{DOC}\cdot \dot{m}_{exh}[T_{DOC} - T_{SCR}] - c_{SCR}[T_{SCR} - T_{amb}] \tag{2.10c}$$

$$\frac{dm_{NOx,tp}}{dt} = \dot{m}_{NOx,e}\left[1 - \eta_{SCR}(T_{SCR},\dot{m}_{exh},C_{NO_2}/C_{NO_x})\right] \tag{2.10d}$$

These equations express the dominant dynamics, which are related to vehicle motion and to thermal and chemical processes in the aftertreatment system. Note that $\dot{m}_{NOx,e}$ as well as exhaust gas temperature and flow, T_{exh} and $\dot{m}_{exh}$, are all a function of $\omega_e, \tau_e, u_{EGR}, u_{VTG}$, and u_{SOI}. The driving torque τ_d is the sum of engine torque τ_e and optional torques corresponding to secondary energy sources. In those cases, an additional state is added: stored battery energy for HEV [27] and power output for WHR system [25]. The optimal powertrain control strategy exploits the energy buffers and EGR–SCR balancing for NO$_x$ emission control. Note that the required vehicle power $W_{vehicle}$ for (part of) the route is available in case of automated driving.

To prevent a possible crash, a minimal d_{veh} is required, where d_0 is the distance at stand still and h is the time headway (in s), which is given by the adopted spacing policy. With respect to vehicle speed, the legal speed limits v_{max} have to be respected and limits to the power sources and the brake restrict the vehicle's acceleration. For the emission constraints, a distinction is made between tailpipe (tp) emission limits for type-approval (TA) and for real-world driving conditions. In the laboratory, a weighted average of cold and hot start emissions (in g/kWh) is determined over the World Harmonized Test Cycle (WHTC) under specified engine operating conditions. This weighted average has to be below the TA limit: $Z_{NOx,TA}$=0.46 [g/kWh]. For the powertrain control problem, this is tackled in [16, 20]. Real-world vehicle emissions have to meet in-service conformity limits, which are based on the work-based window (WBW) calculation method. For windows [0, t_{WBW}] that meet the power threshold ($P_{valid,WBW} \geq 0.20 P_{rated}$) and in which the delivered engine work is equal to the work delivered over the WHTC, the specific emission NO$_{x\_tp,WBW}$ is calculated and put in selected bins. To meet Euro-VI targets, at least 90% of the cumulative frequency distribution $\Phi(NO_{x\_tp,WBW})$ needs to be below 1.5 times the TA limit: so-called conformity factor CF90=1.5.

The described optimal control problem with state constraints has been partly solved. For energy management of HEV, real-time solutions based on Pontryagin's

Minimal Principle (PMP), which give an analytical solution, are widely accepted [4]. Calibration of these strategies is time efficient, since only the optimal (constant) co-state has to be determined. By applying feedback on the battery's state of charge (SOC), performance robustness is enhanced by online co-state adaptation. Also, reference trajectories for the co-state are determined based on predicted W_{veh} [12] and on predicted, stochastic traffic flow information [3]. For the non-predictive HEV case, PMP-based solutions are provided for extensions with gear shifting [21] and aftertreatment [15, 27]. Solving this optimization problem for the extension with tailpipe emissions [Eq.(2.10c)–(2.10d)] is still an open problem; the related co-state dynamics are unstable and have endpoint constraints. Using real-time implementable approximations for the co-state behavior have given good results [26, 27], which are close to the optimal numerical solutions [6, 25]. However, inclusion of ISC limits in the optimization problem is not tackled yet; so far, this was evaluated afterward [16, 20]. For emission management, adaptive co-state control based on tailpipe NO_x emissions is seen to significantly reduce tailpipe emission variations [20]. It is noted that, in case the entire drive cycle is known *a priori*, we can deal with the unstable co-state dynamics and ISC limit. Besides PMP-based solutions, model predictive control (MPC) is applied to (parts of) the sketched optimal control problem. Due to advances in real-time implementation and increasing computing power, engine control applications are within reach, see. e.g. [5, 9, 13].

In conclusion, we propose to extent the adaptive PMP-based method presented in [3] toward integrated energy and emission management. This method can be easily combined with existing low-level and energy management controllers and is expected to set less demanding requirements to control hardware compared to MPC.

2.4 Powertrain Control Development and Testing

Driven by legislation, it has become increasingly important to optimize overall powertrain performance, under a very wide range of operating conditions. As a result, there is an increasing need for evaluation of performance robustness related to fuel consumption and pollutant emissions of future powertrains. An overview of powertrain-related development and testing activities is shown in Fig. 2.5. In this section, first the current status is discussed. Then, an outlook on the potential synergy with ADF-related activities is given in Sect. 2.4.4.

2.4.1 Virtual Development and Verification

To deal with the increased powertrain complexity and exploit the synergy between subsystems, model-based control development is becoming the standard in auto-

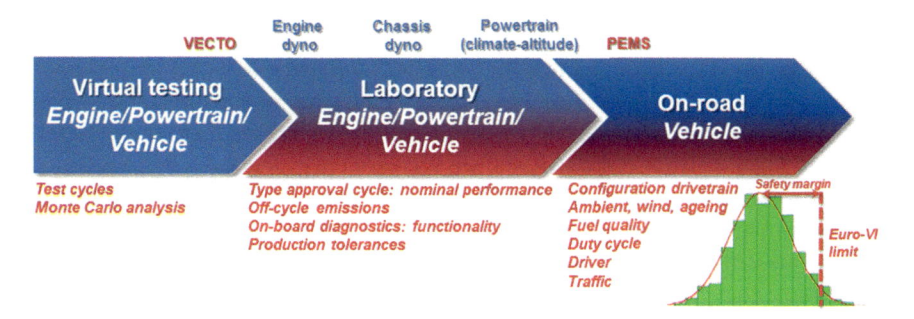

Fig. 2.5 Current status in powertrain development (blue) and verification (red)

motive powertrain development. Physics-based models are developed to design and optimize system layout (sizing, configuration) and control strategies. Also, real-time models can be embedded in controllers to online predict physical quantities, which cannot directly be observed by sensors and powertrain behavior for predictive control applications. From the experience of the authors, up to 30 % reduction of development costs are expected by following a model-based development process for engine-aftertreatment control calibration.

2.4.1.1 Robustness Evaluation

In Sect. 2.2.1, we have seen that real-world driving emissions can vary significantly. To deal with these variations, safety margins are often introduced, as illustrated in Fig. 2.5. However, calibration for lower engine out NO_x emissions typically leads to increased fuel consumption. Virtual testing has been gaining increasing interest to assess various scenarios in a time and cost-efficient way. Monte Carlo analyses are performed to study, e.g., the effect of sensor and actuator uncertainty on system performance [20]. This requires computationally efficient models that accurately describe the powertrain behavior for the studied variations.

To further improve test scenarios, the development of uncertainty models, such as sensor, battery, and catalyst aging models, is important. These models can be used to assess the controller's performance robustness. Also, online model implementation can help to monitor actual aging status and adapt controller settings to further minimize safety margins.

2.4.1.2 Virtual Testing in Legislation

Due to the large number of possible configurations and missions, on-road testing would be an extremely high effort. Therefore, simulation of powertrains' fuel economy is discussed. In Europe, the Vehicle Energy Consumption Calculation Tool

(VECTO) has been developed [8, 22]. This simulation tool predicts tank-to-wheel CO_2 emission and fuel consumption for a selected conventional powertrain, vehicle, and mission. OEMs will provide the necessary input to the simulation models: measured stationary component characteristics, for example, engine, tires, and vehicles air and roll resistance. Control aspects are not explicitly included. Obviously, the quality of the methodology strongly depends on the model accuracy. Note that a similar simulation tool called GEM is used in the USA [8].

First, this tool aims to monitor on-road performance and to help customers selecting the most fuel-efficient solution for their application. However, VECTO will be developed further and is aimed to be used for vehicle labeling in 2018. This includes continuous model validation and improvement with real-world data. A so-called declaration mode is foreseen, in which the vehicle performance is characterized by its CO_2 emission. Unlike in the USA, European CO_2 targets are not set yet for the near future. For 2035, 35% CO_2 reduction compared to Euro-VI vehicles is discussed.

2.4.2 Laboratory Testing

Experimental powertrain testing and validation range from component to vehicle level. In the development phase, test activities focus on system characterization, model development, and control development. For software validation, also hardware-in-the-loop (HIL) testing is done; e.g., OBD software functionality, which is implemented on an ECU, can be efficiently validated by simulating various scenarios. For on-road HD applications, engine dynamometer tests are performed for final validation and for type approval.

2.4.2.1 Robustness Evaluation

Chassis dynamometer testing is still done to validate the overall vehicle performance. With the shift in focus toward real-world emissions, there is a growing need to validate the vehicle's emission performance for various ambient conditions. Climatic altitude chamber testing, with ambient temperature ranging between −45 and 55 [°C] in combination with simulated altitude up to 4000 [m], is currently state of the art. As illustrated in Fig. 2.6, this covers the conditions where OBD and ISC requirements have to be met. This enables time- and cost-efficient pre-validation before on-road testing.

This facility also allows to study worst-case scenarios, for example, cold start, DPF regeneration at low ambient pressure and temperature, and vehicle cooling during hot conditions. With this experimental setup, fuel savings, and CO_2 reduction are determined in a systematic, repeatable, and accurate way on powertrain or vehicle level: ±0.3% FC error for conventional and hybrid drivelines. More details can be found in [10].

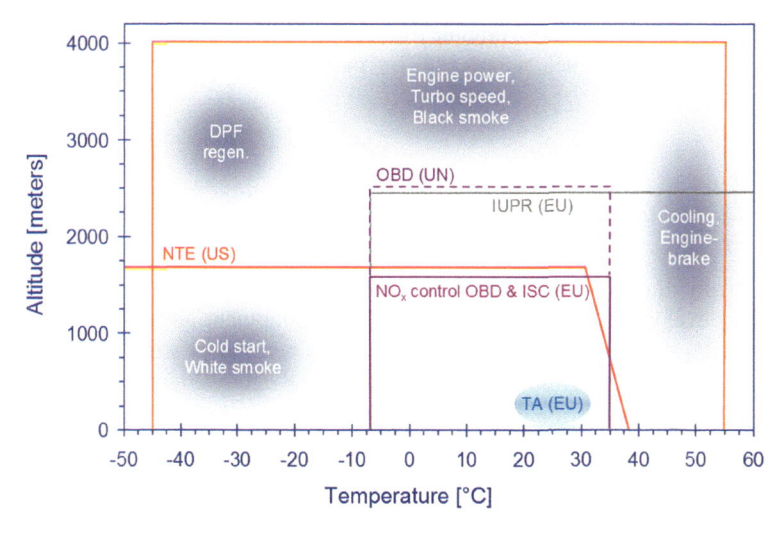

Fig. 2.6 Focus areas for climatic altitude chamber testing (Adapted from [10], © 2010 TNO, used with permission)

2.4.3 On-Road Vehicle Testing

2.4.3.1 Portable Emission Measurement Systems (PEMSs)

To validate their real-world emission performance, vehicles are equipped with Portable Emission Measurement Systems (PEMSs). More precisely, a dedicated measurement system is installed, which online monitors gaseous emissions (NO_x, NO, THC, CO, and CO_2 in g/kWh or g/gCO_2) using variously dedicated analyzers. First, this started as a validation tool to acquire data for governments to determine emission reduction strategies and policies. However, for trucks, this has been mandatory, since the introduction of Euro-VI targets in 2013. The ISC testing has to be done for specific ambient conditions (see Fig. 2.6) and for a specified trip, which consists of a well-defined mix of city–rural–highway driving. Emissions are classified using the work-based window method, as explained in Sect. 2.3.2. Results and more details can be found, e.g., in [23]. Future ISC testing will include particulate number (PN) measurement and will focus more on city driving and cold start.

2.4.3.2 Smart Emission Measurement Systems (SEMSs)

Sensor-based SEMS is used as an effective emission screening tool that provides real-time information on the in-use emission performance for any operating condition and test cycle. These systems use available ECU information from onboard emission sensors and estimated mass air flow and engine torque. Up to now, production type

NO_x, NH_3, and O_2 sensors are available to examine the corresponding emissions and derive CO_2 for a given fuel, see also [23]. Application of CH_4 and soot sensors is foreseen in the future.

2.4.4 Potential Synergy in ADF and Powertrain Development

To fully exploit the ADF potential for real-world CO_2 and pollutant emission reduction, joint development is required. Currently, development and validation of the vehicle's safety and emissions systems are typically done in isolation. Table 2.1 gives an overview of possible combinations of virtual and real testing.

As a first step, installation of PEMS on test vehicles is straightforward. This will generate real-world data of the vehicle's performance. With SEMS, complete fleets can be easily monitored in real-time and huge amounts of data become available of engine, vehicle, and traffic behavior under varying operating conditions. This is a crucial step toward the acceptance of on-road emission trading. Furthermore, it creates the possibility of an *extended development and validation loop*, as illustrated in Fig. 2.7. By coupling this information back, the development of validated traffic and uncertainty (e.g., ambient or aging effects) models and of realistic test scenarios can be significantly accelerated. In [12, 14, 28], examples are found in which engine or chassis dynamometers are coupled with traffic flow models. These applications typically focus on tracking of desired velocity profiles and on FC and are still in a research phase. Also, SEMS and traffic data can be used directly in the control strategy or via a fleet management system to determine route-optimal control settings in order to enhance performance robustness, see Fig. 2.4.

For the development and assessment of integrated energy and emission control strategies, engine, vehicle, and traffic models have to be integrated into a single simulation environment (*software-in-the-loop, SIL*). We strongly believe that this requires adaptation of the existing mixed fidelity models in order to realize an acceptable trade-off between computational efficiency and accuracy. By control function integration on vehicle level, further fuel saving potential is foreseen; e.g., (platoon-) optimal

Table 2.1 Various integration levels of virtual simulation in mixed testing [24]

System level	Software-in-the-loop	Engine-in-the-loop	Vehicle-in-the-loop	On-road
Traffic	Model	Model	Model	Real
Vehicle	Model	Model	Real	Real
Engine	Model	Real	Real	Real
Ambient conditions	Model	Real	Real	Real

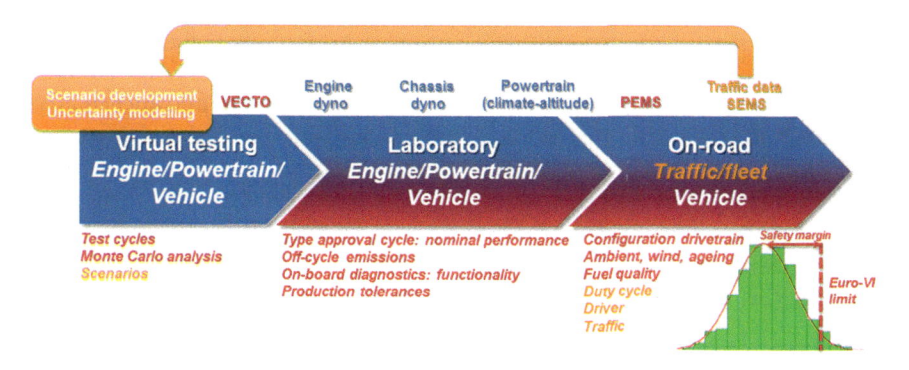

Fig. 2.7 Powertrain development (blue) and verification (red) with extended development and validation loop (orange)

engine settings for following vehicles during truck platooning. For this case, the impact on vehicle cooling requirements also needs attention.

Introduction of ADF systems will also affect certification procedures; different levels of driving automation have to be integrated into on-road certification as well as in simulation tools like VECTO. These tools have to be updated to assess its potential. Moreover, new, representative test cycles for engine certification have to be considered for wide introduction of the highest automation levels.

2.5 Conclusions

In this work, we examined the ADF potential to reduce real-world CO_2 and pollutant emission for on-road HD powertrains. In summary, the main findings are:

- **Robust performance becomes increasingly important**, since attention shifts toward real-world assessment of safety, energy efficiency, and emissions. Simultaneously, the scope for optimization is moving from vehicle level toward traffic and transportation system level;
- **Integration of ADF in powertrain control systems is an important enabler for real-world CO_2 and pollutant emission reduction**. By taking the driver out-of-the-loop, cycle-to-cycle variations and large accelerations can be removed, and buffers can be further exploited by energy and emission trading. This will lead to significantly reduced variations in tailpipe emissions;
- **SEMS is crucial for on-road emission trading**: real-time tailpipe emission data will improve the acceptance and is required for robust control. Currently, studies on the impact of ADF on real-world pollutant emissions are lacking;
- **A general, optimal control framework is introduced, which integrates ADF in integrated energy and emission management strategies**. Solving this opti-

mization problem while explicitly dealing with on-road tailpipe pollutant emission constraints is still an open problem;
- In ADF and powertrain system development, validated traffic and uncertainty models and realistic testing scenarios are missing. Joint on-road vehicle testing will overcome this, since real-world fleet and traffic data becomes available to create an *extended development and validation loop*. Due to improved robustness evaluation, performance can be further enhanced. To deal with the different system levels, smart integration of mixed fidelity models is required for virtual and mixed testing. This will accelerate development and reduce costs.

Acknowledgements The authors want to thank their TNO colleagues Norbert Ligterink, Richard Smokers, Robin Vermeulen, Olaf op den Camp, and Jeroen Ploeg for suggestions and fruitful discussions. Andreea Balau is acknowledged for delivering the real-world duty cycle data.

References

1. ACEA: Commercial vehicles and CO_2. www.acea.be (2010)
2. Al Alam, A., Gattami, A., Johansson, K.H.: An experimental study on the fuel reduction potential of heavy duty vehicle platooning. In: 13th International IEEE Conference on Intelligent Transportation Systems (2010). https://doi.org/10.1109/ITSC.2010.5625054
3. Bouwman, K., Pham, T., Wilkins, S., Hofman, T.: Predictive energy management strategy including traffic flow data for hybrid electric vehicles. IFAC-PapersOnLine **50**(1), 10046–10051 (2017)
4. De Jager, B., Van Keulen, T., Kessels, J.: Optimal Control of Hybrid Vehicles. Advances in Industrial Control. Springer, Berlin (2013)
5. Del Re, L., Allgöwer, F., Glielmo, L., Guardiola, C., Kolmanovsky, I.: Automotive model predictive control: models, methods and applications. Lecture Notes in Control and Information Sciences, vol. 402. Springer, Berlin (2010). https://doi.org/10.1007/978-1-84996-071-7
6. Donkers, M., Van Schijndel, J., Heemels, W., Willems, F.: Optimal control for integrated emission management in diesel engines. Control Eng. Pract. **61**, 206–216 (2017)
7. Eurotransport: LKW-verbrauchswerte von 1966 bis 2014: Immer abwärts. http://www.eurotransport.de/news/lkw-verbrauchswerte-von-1966-bis-2014-immer-abwaerts-6550678.html (In German) (2014)
8. Franco, V., Delgado, O., Muncrief, R.: Heavy-duty vehicle fuel efficiency simulation: a comparison of US and EU tools. Tech. rep, International Council on Clean Transport, ICCT (2015)
9. Gelso, E., Dahl, J.: Air-path control of a heavy-duty EGR-VGT diesel engine. In: 8th IFAC Symposium on Advances in Automotive Control AAC 2016, *IFAC-PapersOnLine*, vol. 49, p. 589595. IFAC, IFAC-PapersOnLine, Norrkping, Sweden (2016). https://doi.org/10.1016/j.ifacol.2016.08.086
10. van Gompel, P., Willems, F., Doosje, E., Vergouwe, M.: Exhaust-gas aftertreatment under extreme conditions : validation in a climatic-altitude chamber. ATZ Auto Technol. **10**(3), 30–35 (2010)
11. Hellström, E., Ivarsson, M., Åslund, J., Nielsen, L.: Look-ahead control for heavy trucks to minimize trip time and fuel consumption. Control Eng. Pract. **17**(2), 245–254 (2009)
12. van Keulen, T., van Mullem, D., de Jager, B., Kessels, J.T., Steinbuch, M.: Design, implementation, and experimental validation of optimal power split control for hybrid electric trucks. Control Eng. Pract. **20**(5), 547–558 (2012). https://doi.org/10.1016/j.conengprac.2012.01.010
13. Kim, Y.W., Van Nieuwstadt, M., Stewart, G., Pekar, J.: Model predictive control of DOC temperature during DPF regeneration. In: SAE World Congress, SAE paper 2014-01-1165. SAE International, Detroit, MI, USA (2014). https://doi.org/10.4271/2014-01-1165

14. Lang, D., Stanger, T., del Re, L.: Opportunities on fuel economy utilizing V2V based drive systems. In: SAE World Congress, SAE Technical Paper 2013-01-0985. SAE International (2013). https://doi.org/10.4271/2013-01-0985
15. Maamria, D., Sciarretta, A., Chaplais, F., Petit, N.: Extension of the ECMS to the real-time energy management of hybrid vehicles with thermal dynamics. In: 4th European Conference on Computational Optimization. Leuven (2016)
16. Mentink, P., van den Nieuwenhof, R., Kupper, F., Willems, F., Kooijman, D.: Robust emission management strategy to meet real-world emission requirements for hd diesel engines. SAE Int. J. Engines **8**(3), 1168–1180 (2015). https://doi.org/10.4271/2015-01-0998
17. Murgovski, N., Egardt, B., Nilsson, M.: Cooperative energy management of automated vehicles. Control Eng. Pract. **57**, 84–98 (2016)
18. NHTSA: EPA and NHTSA adapt standards to reduce green house gas emissions and improve fuel efficiency of medium- and heavy-duty vehicles for model year 2018 and beyond. Publication at https://www.nhtsa.gov/About-NHTSA/Press-Releases/2016/md_hd_cafe_final_rule_08162016 (2016)
19. Ramachandran, S.: Robust Integrated Emission Management strategy for HD diesel engines. Tech. rep., CST 2015.092, Eindhoven University of Technology (2015)
20. Ramachandran, S., Hommen, G., Mentink, P., Seykens, X., Willems, F., Kupper, F.: Robust, cost-optimal and compliant engine and aftertreatment operation using air-path control and tailpipe emission feedback. SAE Int. J. Engines **9**(3), 1662–1673 (2016). https://doi.org/10.4271/2016-01-0961
21. van Reeven, V., Hofman, T., Huisman, R., Steinbuch, M.: Extending energy management in hybrid electric vehicles with explicit control of gear shifting and start-stop. In: Proceedings of the American Control Conference, pp. 521–526. IEEE, Montreal, Canada (2012)
22. Savvidis, D.: Heavy duty vehicle's CO_2 legislation in Europe and VECTO simulation tool. In: 8th Forum on Energy Efficiency in Transport. Mexico City, Mexico (2015)
23. Vermeulen, R., Vonk, W., van Gijlswijk, R., Buskermolen, E.: The Netherlands In-Service emissions testing programme for heavy-duty vehicles 2015-2016 - annual report. Tech. Rep. TNO 2016 R11270, TNO 2016 R11270, TNO, Delft, The Netherlands (2016)
24. Willems, F.: The self-learning powertrain: towards green and smart mobility. Eindhoven University of Technology, ISBN 978-90-386-4252-9 (2017). Inaugural lecture. https://research.tue.nl/files/59039327/Willems_2017.pdf
25. Willems, F., Donkers, M., Kupper, F.: Optimization and Optimal Control in Automotive Systems, Lecture Notes in Control and Information Sciences, vol. 455, chap. Optimal control of diesel engines with waste heat recovery systems, pp. 237–253. Springer, Berlin (2014)
26. Willems, F., Mentink, P., Kupper, F., Van den Eijnden, E.: Integrated emission management for cost optimal EGR-SCR balancing in diesels. In: 7th IFAC Symposium on Advances in Automotive Control AAC 2013, pp. 701–706. IFAC, Tokyo, Japan (2013)
27. Willems, F., Spronkmans, S., Kessels, J.: Integrated powertrain control to meet low CO_2 emissions for a hybrid distribution truck with SCR-deNO$_x$ system. In: ASME 2011 Dynamic Systems & Control Conference, DSCC2011-5981. ASME, Arlington, Virginia (2011). DSCC2011-5981
28. Zhou, J., Schmied, R., Sandalek, A., Kokal, H., del Re, L.: A framework for virtual testing of adas. SAE International Journal of Passenger Cars-Electronic and Electrical Systems **9**(1), 66–73 (2016). https://doi.org/10.4271/2016-01-0049

Chapter 3
Gaining Knowledge on Automated Driving's Safety—The Risk-Free VAAFO Tool

Philipp Junietz, Walther Wachenfeld, Valerij Schönemann, Kai Domhardt, Wadim Tribelhorn and Hermann Winner

Abstract While the technical development of automated driving functions has made significant progress in the past decade, there is still no validation method. An assessment of automated driving before the market launch is difficult because, except for test drives on real roads, there is no information about the behavior of automated vehicles in real traffic. Due to the high level of safety in today's traffic, it is not economically feasible to prove the superiority of automated driving by test drives only, because it would require a testing distance of several million to billion kilometers. To shift testing to simulation or test tracks, information about critical test scenarios is necessary. The Virtual Assessment of Automation in Field Operation (VAAFO) helps to gain more knowledge about automated driving functions while using series vehicles in today's traffic. The new function is tested virtually, and at the same time, information is gained to deduce test scenarios for testing on test tracks and simulation.

P. Junietz (✉) · W. Wachenfeld · V. Schönemann · H. Winner
Institute of Automotive Engineering, Technische Universität Darmstadt,
Otto-Berndt-Straße 2, 64287 Darmstadt, Germany
e-mail: junietz@fzd.tu-darmstadt.de

W. Wachenfeld
e-mail: wachenfeld@fzd.tu-darmstadt.de

V. Schönemann
e-mail: schonemann@fzd.tu-darmstadt.de

H. Winner
e-mail: winner@fzd.tu-darmstadt.de

K. Domhardt · W. Tribelhorn
Technische Universität Darmstadt, Darmstadt, Germany
e-mail: domhardt@stud.tu-darmstadt.de

W. Tribelhorn
e-mail: tribelhorn@stud.tu-darmstadt.de

© Springer International Publishing AG, part of Springer Nature 2019
H. Waschl et al. (eds.), *Control Strategies for Advanced Driver Assistance Systems and Autonomous Driving Functions*, Lecture Notes in Control and Information Sciences 476, https://doi.org/10.1007/978-3-319-91569-2_3

3.1 Motivation

One of the challenges for the release of automated driving is the often-quoted aspect of increased safety in road traffic. A requirement often raised for automated driving is that it should cause less traffic accidents than manually driven vehicles [8]. Thereby, the worst case of reported traffic events is accidents with fatalities. According to the German Federal Statistical Office [19], 377 fatal accidents occurred in 2015 within 237×10^6 km driven on the German Autobahn. At average, an accident with fatalities happens every 628×10^6 km. Winner and Wachenfeld [24, 31] consider a distance factor of around 10 for statistical proof of an increased safety. Theoretically, a test distance of around 6×10^9 km has to be driven. Each statistical safety verification needs to be restarted once the system is modified. These distances are not drivable in an economic way based on regular real-world test driving if compared to today's distances driven for testing [3].

Consequently, the release of automated driving can become an actual trap if such a statistical testing method is solely applied. Other approaches have to be considered in order to decrease the testing periods and achieve satisfying results when demonstrating the trustworthiness of automated driving.

Beside the statistic approaches for solving the challenges of testing mentioned in [24, 31], other techniques can be considered to shorten the verification phase, especially when a system is modified. Since only a small amount of the driven distance is supposed to be safety critical, a scenario-based identification of critical traffic situations can be applied in order to develop reasonable test cases for automated driving. A scenario catalog is developed by characterizing relevant traffic settings. According to Ulbrich et al. [20], a scenario is defined as a composition of scenes with actions and events as well as goals and values in between. A variety of parameters such as road infrastructure, weather conditions, and different types of road users also describe a scenario. Once the necessary scenarios are found, test cases can be generated. However, a huge number of possible scenarios exists but not all of them are safety critical. A major issue is the identification of the most relevant scenarios. A tool is required which helps to identify these scenarios. A new concept for identification introduced by Wachenfeld and Winner [23] is the Virtual Assessment of Automation in Field Operation (VAAFO) which extracts relevant cases from a huge amount of kilometers driven in the random nature of the real world.

The following content: In Sect. 3.2, the VAAFO tool is described in detail. In Sect. 3.3, thoughts on the extrapolation of the scenario identification are laid out with the goal to handle the long relevant distance derived from statistics. Possible applications of the VAAFO tool are discussed in Sect. 3.4.

3.2 Virtual Assessment of Automation in Field Operation

Different tools for the assessment of automation exist, as depicted in Fig. 3.1. Besides real driving, artificial methods, e.g., tests on proving grounds or virtual methods, e.g., simulation exists. All other tools besides real driving suffer the challenge of validity.

		Environment		
		virtual	artificial	real
Vehicle	virtual	SiL		-
	artificial			
	real	-		real driving

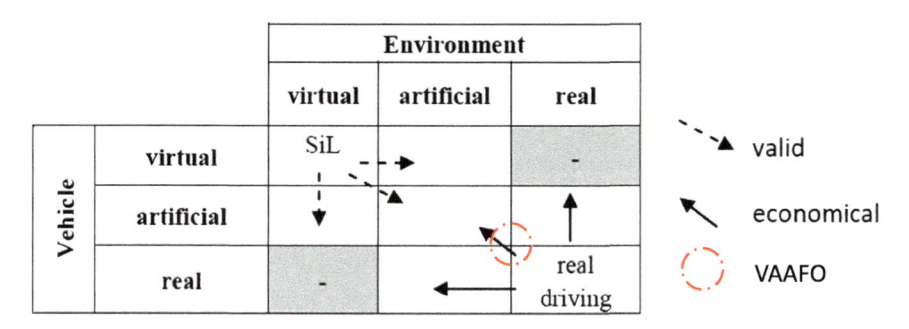

- - → valid
← economical
◯ VAAFO

Fig. 3.1 Classification of testing tools for testing automated vehicles [24] (Reproduced from Open Access Article)

The validity of the test case for being relevant as well as the validity of the tool itself are questionable. However, these tools gain economic benefits. Consequently, the VAAFO approach tries to stay as close to real driving as possible while still enabling the economical coverage of reality and thereby reducing additional risks to zero.

This section introduces VAAFO in general and gives a systematic overview of the use cases from the VAAFO perspective. Use cases of a testing function are different ones than use cases for driving functions as will be explained. The basic VAAFO architecture is explained, highlighting the need for two main components of the concept. Both components, the "Correction of Environment Perception" as well as the "Challenge of Initialization", will then be discussed in more detail.

3.2.1 Introduction

From a functional point of view, automated driving can be decomposed in three major levels: The automated driving function senses, thinks, and acts. All possible sources of risks in terms of safety can similarly be assigned to one of these three levels, as presented in [24]. Level one describes causes that happen within the information perception phase. The absence of sensor information leads to an absence of data needed to be processed. For example, an object may be covered or contain undetectable characteristics. Level two classifies all errors that lie within the information processing such as the application of nowadays algorithms. Level three categorizes all causes of accidents that occur after the decision is made, due to improper control of the vehicle movement. Table 3.1 shows such a categorization of accident causes for automated driving.

With VAAFO, the automation algorithm is executed virtually during normal driving. On the one hand, the VAAFO tool uses the decomposition of the vehicle automation to prevent additional risk by only simulating the desired action (level three/act) of the automation. On the other hand, it uses the real sensors (level one/sense) as

Table 3.1 Classification of accident causes for automated driving

System	Category	Causes
Automated vehicle	Sense	e.g., object is covered or contains undetectable characteristics, poor environment conditions, broken, soiled or misplaced sensor, too short observation time, etc.
	Think	e.g., misunderstanding of situation or wrong anticipation of traffic environment, wrong behavior strategy, etc.
	Act	e.g., defective or uncalibrated actuators, failure of necessary vehicle components, etc.

well as the real processing hardware (level two/think) to stay as close to the real automated driving function as possible.

The VAAFO tool requires the following hardware:

- The basis is a series vehicles that is driven by a human (SAE level 0), assisted by advanced driver assistance systems (SAE level 1), or partially automated (SAE level 2) [16].
- This series vehicle is equipped with sensors suitable for higher automated driving (SAE level 3, 4, and 5, called AD3+). Those sense the real environment. (Operating VAAFO with less powerful equipment and reduced functionality is also possible, see Sect. 3.4.1.)
- Additionally, the microprocessors and respective processing algorithms are installed. These algorithms process the data coming from real sensors similar like the ones later used for AD3+.

The VAAFO tool is applied onto the resulting perceived environment representation, as the following section explains.

3.2.2 System Architecture

The VAAFO tool is composed of three functional modules:

1. Supervision and correction of the perceived environment representation;
2. Simulation of the automated short-term behavior in the perceived environment representation;
3. Safety assessment of the series vehicle reference behavior vs. the automation behavior. Both assessments are executed in the corrected environment representation.

The simulation, correction, and assessment functionalities consist of multiple sub-modules. The most important steps are:

The correction module modifies the perceived environment of the reference scenario based on false positive and false negative indicators concerning the existence

3 Gaining Knowledge on Automated Driving's …   51

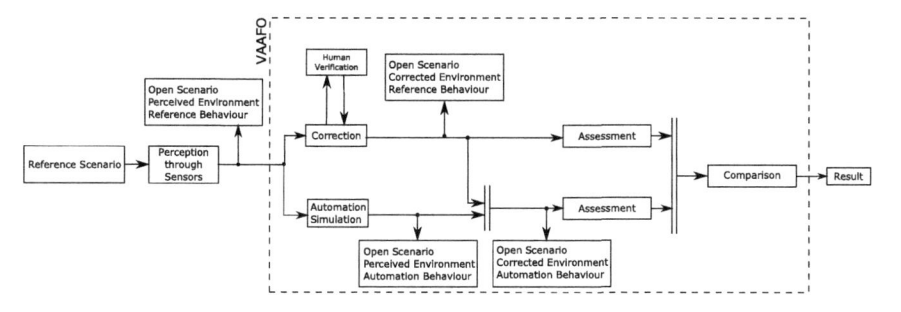

Fig. 3.2 Process architecture of a VAAFO implementation

of object representation. The result of the modifications is a corrected environment with reduced false positives or false negatives that are of relevance for the driving task. A more detailed explanation can be found in Sect. 3.2.4.

The simulation module uses the perceived environment representation and creates the (short-term) behavior of the automation within this environment representation. The automation acts virtually in the real perceived environment representation. This is further described in Sect. 3.2.5.

In order to assess the automation's behavior in the real environment, the relevant parts of the results from correction and simulation are simply joined for further assessment. For the assessment module, the simplest measure is whether a real collision has occurred or not. Further, multiple metrics for preventive or active safety [30] can be implemented. Those can range from the safety in the current motion state (TTC, constant reserve time, etc.) [30] to motion prediction models (physics-based, maneuver-based, and interaction-aware) [13].

VAAFO is intended to be applied while the reference vehicle is driving. Therefore, the implementation needs to be able to run as a soft real-time system. As opposed to hard real-time, soft real-time does not result in a critical failure if the real-time requirement is not met.

The system has three separate interfaces, which are both in- and output, and an additional one for the environmental correction. The input interface provides the world representation. For the world representation, an adapted OpenScenario format [21] is used. Within the process, the correction functionality has the option to use a human user as an additional source of information and ask him to help interpret inconclusive scenarios. The output consists of the result of the assessment as well as the perceived environment (Fig. 3.2).

3.2.3 Use Cases

This section describes the use cases for the testing functionality of VAAFO. The testing functionality compares virtual automation behavior with real, manual, or semiautomated reference behavior in the perceived and simulated environment.

A VAAFO use case is a traffic scenario in which VAAFO is activated (see Sect. 3.2.5.2 for activation conditions). Every use case for the VAAFO method consists of four related scenarios, as indicated by the four columns of Fig. 3.3. Two scenarios are within the real environment. The real environment is always unknown because it can never be detected in total. The other two scenarios are in the perceived environment. This environment is perceived by AD3+ sensors and algorithms and known from the beginning in real-time. In both environments, there is one scenario containing the reference behavior of the driver or assisting functions and one with the behavior of the automation. Additional information about the real environment can be added at a later time when additional information is known or estimated. There is no real-time constraint. However, the information about the real environment is never complete.

Relevant for the VAAFO approach is those use cases that do not result in a real conflict (accident) for the reference (real) behavior within the real environment. It is

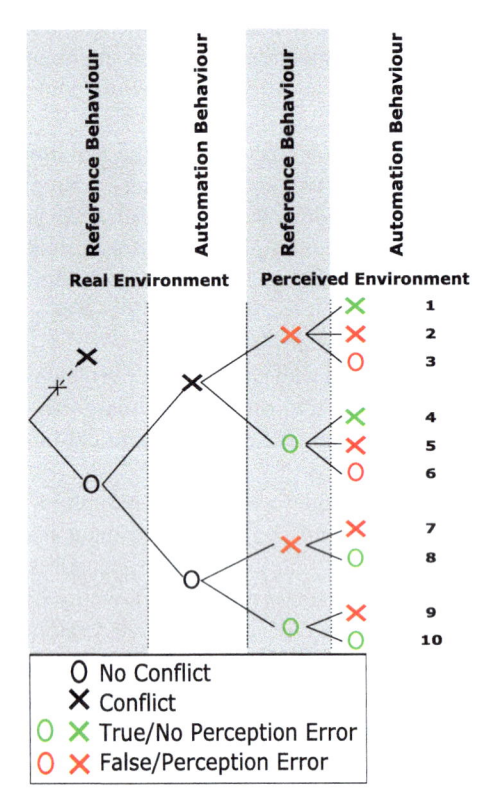

Fig. 3.3 Conflict perception for VAAFO use cases: The perceived environment can possibly contain false information. Therefore, the reference behavior may appear to result in a conflict (False/X). The same applies to the automation: Although the automation would drive conflict-free in reality, the perceived environment representation indicates a conflict. Further, if there were a conflict of automation in the real environment, it is also possible that no conflict is perceived (False/0)

assumed that these use cases are recorded by police. The use case tree in Fig. 3.3 is terminated for this upper branch.

For the other lower branch, the objective is to determine whether or not the behavior of the automation would have resulted in a conflict (X) (virtual accident) within the real environment.

Figure 3.3 classifies ten use cases for the VAAFO method. The most important use cases occur whenever the automation behavior results in a conflict, but this is not perceived correctly. Without a correct assessment, these use cases can result in the approval of an insufficiently safe automation, as explained in the following. These are the use cases 3 and 6.

As an example, use case number 6 is explained: A reference vehicle drove a series of maneuvers without getting involved in an accident (0). The automation, however, would have gotten into an accident (X). In the perceived environment, both the reference (T0) and automation (F0) appear as accident-free. The assessment of the reference behavior would yield a correct result. However, there is a critical discrepancy between what happened to the automation in the real and in the perceived environment. Therefore, the assessment of the automation would yield false results. This use case can be caused by differing trajectories between the reference vehicle and the automation and an error in the perceived environment that places a traffic object in the wrong position.

It is the objective of the VAAFO method to deduce, from the given perceived scenario pair, the corresponding true scenario pair in the real environment. In order to bridge this informational gap, identification and ideally a correction of the perceived environment are necessary for any useful assessment of an automation using the VAAFO method. Possible approaches are discussed in the following.

3.2.4 Correction of Environment Perception

State-of-the-art environment perception uses ultrasonic [14] sensors, automotive radar [27], automotive lidar [11], and automotive camera [9]. The data of these sensors are fused to an environment representation in different ways [5]. A commonly used [6] representation is the object-based representation of ego as well as other objects. This object-based representation suffers [6] state as well as existence uncertainties (uncertainty about classification is neglected). Dietmayer et al. [6] describe the confirmation of an object as really existing by:

> [...] current systems usually determine the existence of an object based on a heuristic quality criterion $q(x)$ of an object hypothesis. An object is considered confirmed if the quality criterion exceeds a sensor and application dependent threshold θ.

For a driving function, the quality criterion for online decisions can only be derived based on the measurements from the past, which are processable in real time and collected by the sensors connected to the car. (It is assumed that C2X will not be available for safety critical functions.) The driving function needs to come to a decision and act despite these restrictions. These restrictions of a driving function are avoided by the VAAFO tool, as the real-world action is not necessary. Without these restrictions, additional information sources are accessible for the VAAFO tool to correct the environment perception of the automation in an automated way. In principle, three approaches and a combination of them exist:

1. Measurements from future time steps get accessible as the hard real-time constraint does not exist. A retrospective evaluation is possible.
2. Due to the same reason, the time budget for post-processing steps increases and enables the calculation of more advanced algorithms.
3. Additional information sources besides the environment sensors are available due to the human driver (SAE level 0 and 1 [16]) and his or her interventions (SAE level 1 and 2 [16]). The human driver serves as an indirect environment sensor assessor.

The first two approaches still suffer the uncertainties of state and existence as it can never be guaranteed that the environment perception (sensors and algorithms) will detect the object at a later driving situation. Nevertheless, due to a systematic improvement, the potential of reducing false negative as well as false positive identifications exists. It depends on special sensors as well as algorithms to prove and quantify the improvement.

The third approach tries to access the human sensors by a model-based approach. It detects situations where the human behavior cannot be explained with the sensor information. The accident-free drive of a human proves that his or her perception was free of false positive (FP) and false negative (FN) detections that result in an accident. In general, the perception of a human is not free of FP and FN [15], but the controllability of the human himself and the other drivers in his or her surrounding together correct many of these mistakes.[1]

The model-based approach to access the human as a sensor is separated in three sub-elements (see Fig. 3.4). The correction of the environment representation is indicator-driven. The main goal is to identify relevant FP and crucial FN objects, as has been explained by the use cases in the section before. Therefore, the human behavior is used as an extra information source. In comparison to the environment perception, FN and FP indicators are derived.

The FP indicator is derived by the overlapping of perceived objects and an ego-state adaptive area in the surrounding of the ego-vehicle, as well as their length of stay within this area. In Fig. 3.4, this is called the FP-Area check. If certain thresholds are exceeded, an FP indicator is raised for the respective object. The logic behind this is the assumption of a risk-minimizing behavior of a human driver.

The FN indicator is derived based on an adapted wish-based approach following the lane change model of Schakel et al. [17], which estimates the driver's wish to conduct a lane change. The presented model is extended to longitudinal desires to

---

[1] It is assumed that a relevant accident of other traffic users due to the ego behavior will be perceived. A combination with the eCall could help to achieve this.

3 Gaining Knowledge on Automated Driving's ...

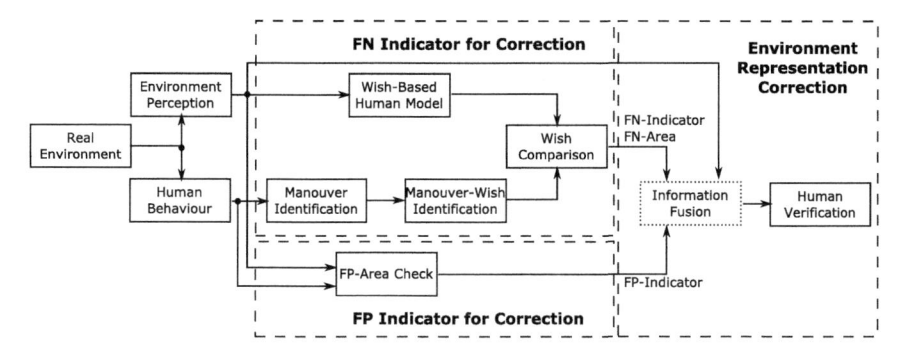

Fig. 3.4 Generalized methodology for the correction of the environment perception

have longitudinal as well as lateral wishes for driving on the German Autobahn.[2] The wish-based human model calculates desires by evaluating the perceived environment representation. Additionally, the inverse wish model calculates the corresponding maneuver wish based on the identified maneuver the human driver chooses by using his (mainly vision-based) perception. By comparing both wishes over time, the indicator for possible FN objects is derived. The instantiation of an object based on the FN indicator is currently not possible. The model-based approach delivers areas where the indicator expects an FN object, as depicted in Fig. 3.5. Each of the indicators depicted in Fig. 3.5 is based on the two wish models. As an example, FN indicator one is active if the wish model asks for a lane change left (W-LC-L) and the human model has not found this lane change (H-LC-0). The indicator assumes that the perception has missed to perceive an object in the areas L1 and/or L2. The other indicators work the same way as shown in Fig. 3.5.

Theoretically, the area could be used similarly like the FP indicator to adapt the heuristic quality criterion $q(x)$ of an object hypothesis, as described above. Although the information base is extended, the uncertainties, even if reduced, still exist.

Due to that fact, an iterative verification step is introduced. A large number of human labeling tasks are necessary, though the indicator approach enables the efficiency improvement of the labeling tasks. First of all, it is not necessary to label every kilometer that is recorded. Due to the indicators, the interesting segments are already highlighted. Moreover, the indicators can directly be translated in a question that the labeler needs to answer. For the FP indicators, it should be asked whether a highlighted object in the video data really is a relevant object. For the FN indicators, at least two segments of the FN area at the beginning and the end of the uncertain time frame are presented and the labeler is asked to mark an object if existing. This simplification of the labeling task might enable gamification approaches with a crowdsourcing approach to handle the big amount of data that is going to be analyzed (see Sect. 3.4.4). The described indicator-based approach is implemented in simulation (Fig. 3.6).

---

[2]This model is mainly the result of the Master thesis of [32].

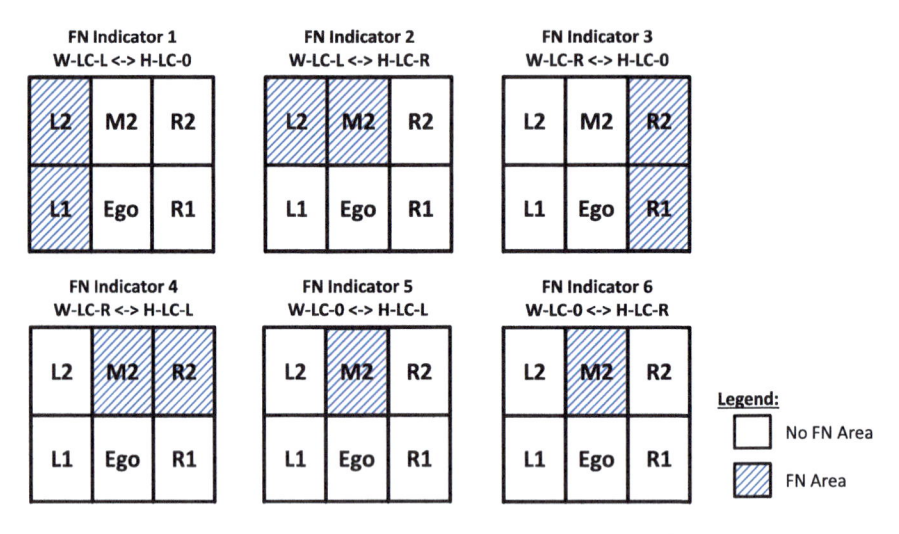

Fig. 3.5 FN areas based on different FN indicators; W-LC-L means *Wish Lane Change Left*; H-WC-L *means Human Lane Change Left*; The ego-vehicle starts always in the middle lane position 1. Left and right lanes are also separated into two positions

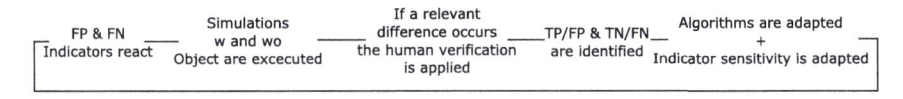

Fig. 3.6 Process of environment correction

3.2.5 Simulation Environment

As the automation within the VAAFO tool is not able to act on real actuators, a virtual simulation of the environment is necessary. By implementing a simulation environment, the automation is able to act on the virtual actuators and thereby changes the course of a virtual scenario. To implement the simulation environment, certain simulation models as well as their starting, ending, and initialization conditions need to be defined.

3.2.5.1 Simulation Models

In principle, the virtual simulation comprises static and dynamic elements. The static elements contain the road and other non-acting objects like traffic signs. These elements will be represented as perceived by the real sensors and their post-processing. These static elements are not influenced by any changes of the course of the virtual scenario: the road course, for example, does not change. Dynamic elements, however, are affected by the ego-vehicle's behavior. Consequently, the dynamic elements are separated into the ego-vehicle as well as object–vehicles (or dynamic objects in general). Do they and how do they behave within the virtual simulation?

The behavior of the ego-vehicle is defined by the algorithms of the automated vehicle. Thus, there is no model to be defined. The behavior of the object–vehicles, however, needs to be actively modeled. From a qualitative perspective, the traffic participants' behavior may be described by the three levels of Donges [7]: navigation tasks, guidance tasks, and stabilization or control tasks. These levels are associated with a necessary time budget they are executed in. Donges estimates that the navigation task reaches from several hours up to a few minutes. The guidance depends strongly on the complexity of the task but requires between a minute and several seconds. For the control task, Donges reports several a delay time of 100 ms. Due to the reason that the virtual simulation tries to stay as close to the real environment as possible, the simulation time should be limited. It is assumed that the behavior model of the object–vehicles can read the navigation and long-term guidance results from the perceived object behavior in reality. On the other hand, the control tasks of the object–vehicle need to be simulated alongside the trajectory of the real behavior. Different control behaviors of the object–vehicles will likely lead to different results of the simulation. Let us assume the case of a passing faster vehicle on a highway. For simulation, it is assumed that this object–vehicle tries to pass as well. If the automation lets it pass, the recorded trajectory will just be repeated in simulation. If the automation pulls out despite the fast-approaching object–vehicle, it reacts during simulation based on a control level driver model. Depending on parameterization, the reaction may happen in time to prevent an accident. Worst-case assumption (e.g., heavily distracted driver) will most likely result in a crash. Best case assumption (attentive driver), however, will almost never cause any accidents. To solve this dilemma, the simulation of an intermediate driving skill is proposed (e.g., by an assumption of the worst 5% of drivers).

Obviously, the longer this closed-loop reactive virtual simulation is executed, the bigger is the difference between virtuality and reality. Starting, ending, and initialization conditions are necessary.

3.2.5.2 Starting, Ending, and Initialization Conditions

As the errors of automation that the VAAFO tool is looking for are unknown, the tool can be seen as a search engine without knowing what the actual case of interest is. Therefore, the goal for the VAAFO application should be to cover the whole distances driven. The definition of any kind of relevance indicator suffers the risk of false neglection. Thus, the strategy for starting and ending the simulation is to come to a complete coverage by just starting in equidistant time steps, as depicted in Fig. 3.7. The figure illustrates that simulations are run in parallel and started at equidistant points in time, every 0.5 s for example. The total runtime is assumed as 2 s. It is assumed that simulation with a longer runtime would not be valid because the behavior of the other traffic participants could vary depending on the simulated vehicle. This could only be handled with an advanced driver behavior model. Until

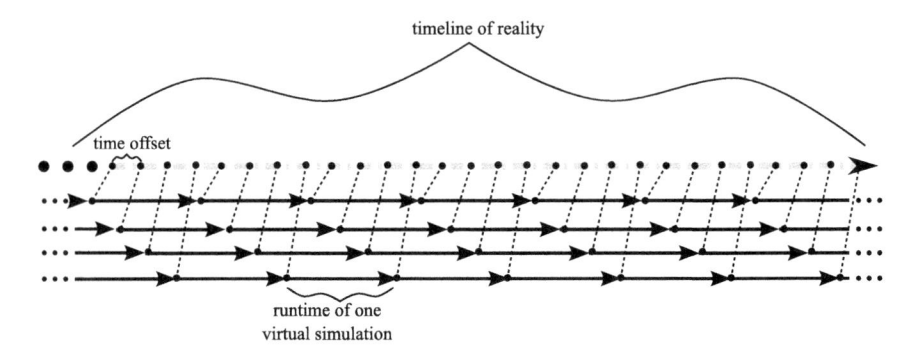

Fig. 3.7 Starting and ending concept of parallel simulations relative to the reality timeline; the equidistant starting times are marked in the top layer. Below, four parallel simulations are executed because the runtime of one simulation is four-times longer then the time offset

now, the authors believe that a short simulation time is sufficient but it should be investigated further as soon as VAAFO is implemented in real traffic.

Through this concept, start and end conditions are defined. However, the question is how to properly initialize all relevant models of object–vehicles and the ego-vehicle automation. Because models as well as the automation consist of processing steps that require the knowledge of past processing steps (finite state machine or simple PID-controllers), the initialization raises this requirement for automation development:

If the VAAFO tool will be applied for testing and validation, the automation must be implemented in a way so that it can be activated at every point in time. This requirement is already fulfilled for an emergency automation but not necessarily for a full automation that drives automated from start to end. Especially due to the reason that tactical driving due to uncertainties might end up in different working points of the full automation, this requirement raises a challenge for the application of VAAFO [12].

3.3 Scenario Identification

On the one hand, the VAAFO tool can address the statistic approval. It might cover travel distances of millions of kilometers a year by application into series vehicles without introducing higher risks to road traffic participants. On the other hand, it also helps to identify new scenarios for the scenario-based approval strategy (see Sect. 3.1). To collect known scenarios for this approval strategy, different data sources are available. Wachenfeld et al. [22] name various data sources such as test drives, naturalistic driving studies (NDS), field operational tests (FOT), and accident databases. The disadvantage of these data is that they describe driving with a conventional vehicle. For existing databases, outdated sensor hardware is often used, which does not represent the environment in a way that suits state-of-the-art or even future vehicles.

Accident databases only collect accidents with state-of-the-art vehicles, which are not necessarily relevant for an automated vehicle. Furthermore, they do not necessarily pose a representative selection of today's traffic accidents. Thus, most of the existing data do not cover all information that might be necessary for identifying relevant scenarios. A knowledge gap currently exists. However, it is the authors' conviction that this data will help with the challenge of gaining the necessary knowledge about the relevant scenarios.

Nevertheless, the questions should be discussed how to acquire necessary additional information to fill this knowledge gap and which test distance we need to cover in order to gain enough information about automated driving.

As a worst-case assumption, we need at least as large recorded and analyzed automated driving distances as the statistic approval strategy demands. Collecting real-world data for mileage in this order of magnitude is not practical, as the data storage of the raw sensor data requires enormous hardware and the effort for generating a ground truth by manual labeling (see Sect. 3.4.4) is enormous.

Instead, VAAFO can be used to identify situations. Based on today's data [22], a basic scenario setup is made. However, there is no way of proving that all eventualities (or even all statistically relevant ones) are covered. Whenever VAAFO detects a relevant situation, it is compared to the existing scenario database. If no match is found, the situation is transferred to a scenario and saved in the database as a new, relevant scenario.

With increasing VAAFO mileage, more and more new scenarios are saved into the database. Assuming a limited number of relevant scenarios, the rate of new surprising situations decreases (comp. Winner [29]). Assuming, e.g., an exponential decrease of this rate, the number of missed situations linked to the remaining approval risk might even be estimated (see Fig. 3.8).

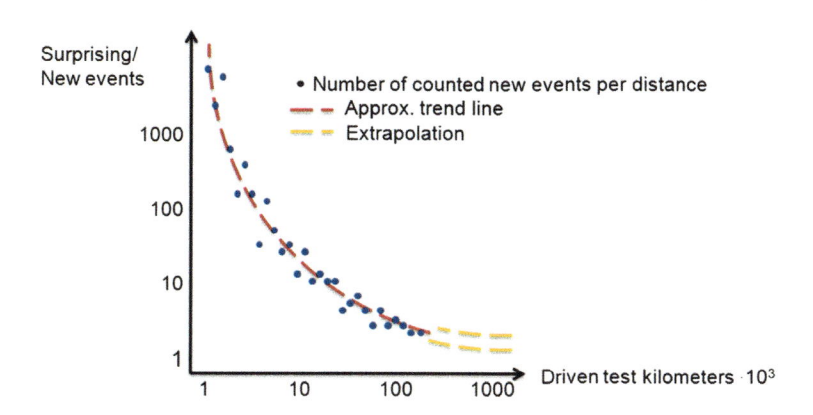

Fig. 3.8 New situations (surprises) during real-world testing [28]

3.4 Application Ideas

On the one hand, the motivation for an application of the VAAFO concept is given by the safety assessment challenge of automated driving. Nevertheless, the development and application of VAAFO are connected with additional costs that are not directly apparent to the customer. Thus, different ideas for application of the VAAFO tool are discussed in the following with their pros and cons.

3.4.1 VAAFO Use with State-of-the-Art Vehicles

In general, VAAFO can be used during all levels of automation, no matter if any advanced driver assistance systems (ADAS) are activated or not. During driver only driving, VAAFO still observes the environment and compares the driver's actions to the system's virtual trajectory planner. However, as in state-of-the-art systems, a driver is always part of the control concept, and thus VAAFO can be used. Winner [25] describes the evaluation of automated driving functions with a Triangle (see Fig. 3.8). State-of-the-art systems approach the target of level 3 automated driving from three directions: low-speed scenarios [(e.g., automated parking (AP)], high-risk scenarios [(e.g., collision avoidance by braking (CA-B)], or in simple scenarios (e.g., highway pilot). In the last two categories, the functions can be tested using VAAFO, as described in the following (Fig. 3.9).

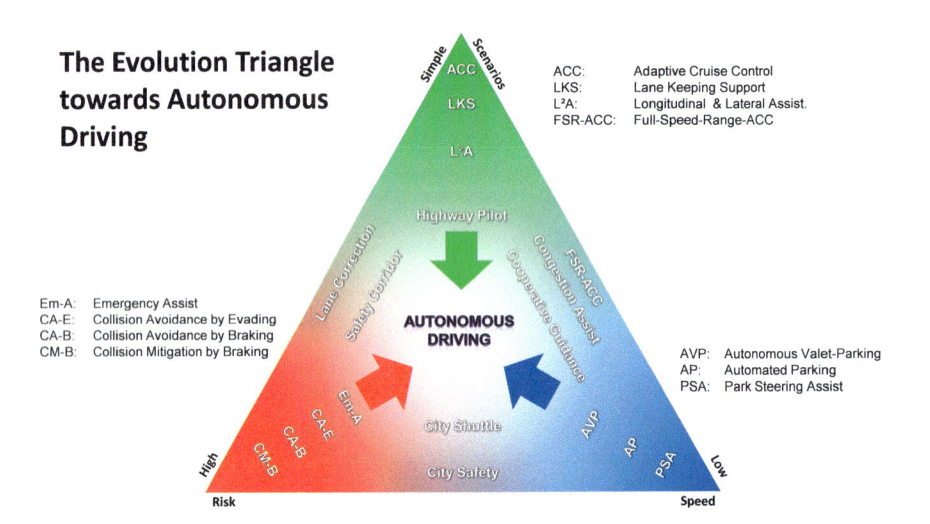

Fig. 3.9 Triangle of automated driving development, from [25] © 2016 Springer, reproduced with permission

For the high-risk functions (first direction), let us focus on (CA-B). State-of-the-art CA-B systems only execute an emergency maneuver if it is unlikely or impossible to prevent an accident without hard braking. Before the automated braking, the driver is warned by visual or audible signals with increasing intensity [26]. In doing so, false activation is prevented for the costs of less correct activations. In most cases, when CA-B decides not to intervene but when the situation requires braking, the human driver will brake. These cases collected during real-world driving are usually unknown to the developer as there is no data storage and no selection of the relevant sensor data available. Only in those cases where neither the driver nor the CA-B brake, data can be available because then the result is a crash. Or in other words: Knowledge about the function is only gained whenever an accident occurs. This is undesirable, not only because the goal is to reduce accident numbers in the future, but also because accidents are very rare.

With additional VAAFO use, data can be saved whenever the driver decides to act differently from the system or, in the case of CA-B, whenever the driver executes an emergency brake and the system was not activated. In addition to that, different parameterization of the function could be used in the virtual environment and compared with the function in the real environment and the driver's action.

For the simple scenario's functions (second direction), state-of-the-art partially automated driving systems combine an adaptive cruise control with a lane keeping assist. During normal assisted driving, human intervention is not necessary. Nevertheless, the driver is obligated to observe the traffic and the surroundings and intervene whenever necessary. The procedure with VAAFO would be quite similar to CA-B only that all scenarios involving human intervention are examined. The concept is comparable to tests with highly automated functions and an additional professional test driver, only that in this case the test is done during normal operation. Caution: The actual automated function needs to be approved and communicated as SAE level 2 or less, thus a partially automated system.

Of course, tests can be done in an Hardware-in-the-Loop- or Software-in-the-Loop-environment as well, but the benefit of virtual testing in field operation is the validity of the environment representation and testing in real traffic, which might differ from a simulation database. Especially the advantages of testing in real traffic must be pointed out. All simulation or test field approaches have the disadvantage that only known scenarios are tested, while the real-world approach offers an unlimited number of unknown "test" scenarios.

3.4.2 Approval by "Seed Automation"

In Sect. 3.1, the "approval trap" for highly automated driving is explained. The high requirements for a statistic approval are mainly based on the complexity of the traffic (and therefore high number of the possible scenarios) and the relatively small number of severe accidents. However, to reduce complexity, a function may be limited to special known areas. Thus, the function development is completed faster if the function is only available in a small part of the road network (e.g., 200 km motorway). One famous example is Volvo's "Drive Me" program [4], where 100 cars will be equipped with highly automated driving functions to be able to drive on the city ring around Gothenburg. On other roads, the function stays disabled. However, the sensor equipment for highly automated driving is available in the car, so VAAFO can be used in all driving conditions. Where the function is restricted because there is no approval for this road section, it is enabled virtually (without actuator access) and tested in field operation. Through virtual operation, more and more testing kilometers are acquired. After a certain amount of accident-free travel distances, an additional section of the road network is cleared and the function approved for this section. The amount of required testing distance depends on the annual mileage and annual accidents on this section (see Wachenfeld and Winner [24] with numbers for the German Autobahn network). This so-called approval by "seed automation" enables the equipment of serial vehicles with state-of-the-art sensor components because there is an actual benefit for the driver (driving automated on designated roads). At the same time, the driver is helping with further approval by driving his vehicle.

3.4.3 Vehicle Equipment for Future Applications

As mentioned before, the vehicle must be equipped with state-of-the-art surrounding sensors for optimal VAAFO performance. With lower sensor quality, more perception errors than necessary are detected. Obviously, VAAFO works best with an additional wireless data connection. Otherwise, the results of the virtual test and identified critical scenarios cannot be sent directly and must be stored in the vehicle, e.g., until the next service. In addition to that, new functions can be uploaded to the vehicle and approved virtually before enabling them in the real world. The new or updated function is sent to the VAAFO tool and assessed virtually in public driving.

Not only in Germany, the privacy of customers is a big issue. Hence, car manufacturers struggle with the introduction of data recording in serial cars. In addition, VAAFO works best when the sensor setup is at least close to the one of a future automated car. For most manufacturers, it would not be economically feasible to equip their series cars with high-end environment perception sensors without selling a function to the customer. The "Seed Automation" (see Sect. 3.4.2) could help with customer acceptance and willingness to pay for the function. Another method is to introduce VAAFO in premium cars such as the Tesla Model S. All Tesla Model S's are equipped with the same sensor setup regardless of the function that was paid for. In addition to that, an Internet connection is available for software updates but also to send data to a backend. In a talk for MIT Technology Review, Sterling Anderson [18], Director of Tesla's Autopilot Program stated:

> Since introducing this hardware 18 months ago we've accrued 780 million miles. [..] We can use all of that data on our servers to look for how people are using our cars and how we can improve things.

However, it is questionable that Tesla really transfers raw sensor data of 780 million miles with their wireless data interface. When acquiring this amount of data, the challenge on how to filter the relevant data is raised. One possibility would be to transfer object lists instead of raw data. VAAFO pursues a different approach, instead. The relevancy of the data is filtered inside the vehicle and only interesting situations are saved with a high level of detail.

3.4.4 VAAFO on a Dedicated Fleet—A New Business Model

The authors of this paper assume, as has been explained in Sect. 3.2.4 on representation correction, that automated correction may not totally work without manual labeling. To solve this, different approaches are feasible. The most intuitive solution is to pay humans for labeling. As labeling is a time-extensive task, this is typically done in low-wage countries.

A different approach is gamification. The labeling task is split up into small parts and short tasks, so a single labeling task takes only a couple of seconds. Examples for gamification are the replacement of advertisements in Smartphone apps [2], Google's reCAPTCHA [1] where the user has to pass a small test on Web sites to protect them from automated queries, or Amazon's Mechanical Turk offer [10] where everyone can do small tasks for little payment.

Correspondingly, the VAAFO concept can be combined with the gamification approach. Road transportation systems, like long-distance coaches for highways or taxis and city busses for inner-city scenarios, are equipped with the required sensors and processing setup for VAAFO. The business idea would be to offer passengers in these vehicles the possibility to get a refund on the ticket price for their ride if they do some of the labeling tasks. To make the tasks more interesting, situations from the actual ride could be labeled. The funding for equipment and refunding would come from companies testing and developing their automated driving function.

3.5 Conclusion

While the technical development of automated driving functions made significant progress in the past decade, there is still no validation method. With state-of-the-art methods, the safety approval is not economically feasible. The core question is how to prove that the introduction of automated driving will raise the safety level on today's roads (see Sect. 3.1). As severe accidents are rare, this challenge is even more difficult to achieve. Very rare traffic scenarios that might cause problems with automated driving are relevant.

A major challenge is the lack of knowledge about these scenarios. It cannot be deduced from today's traffic. VAAFO is one approach to further close this knowledge gap. Instead of testing automated driving in an unsafe environment, the new functionality is tested virtually in real traffic while the vehicle is driven by a human driver. The decisions of the driver contain additional information about the environment and are used as further input to compare it with the information from the perception sensors.

An advantage compared to testing by simulation or on test tracks is that the "proving ground" covers the future field of application: the real traffic. A statistical approval or an approval by "Seed Automation" (see Sect. 3.4.2) is possible in general.

In this paper, the technical challenges of the VAAFO method and different ways of implementing VAAFO are described. The implementation in a testing vehicle is still in progress. The concepts propose a method to gain further knowledge about the virtual behavior of automated driving and about unknown traffic situations, even ahead of highly automated driving with state-of-the-art vehicles.

It has to be pointed out that VAAFO cannot be the only source to close the knowledge gap. Many different methods have to work together to acquire knowledge with the common goal of approving the safety of automated driving.

References

1. Amazon.com Inc.: Amazon mechanical turk. https://www.mturk.com/mturk/welcome
2. Andrulis, J., Kondermann, D.: Pallas Ludens: We inject human intelligence precisely where automation fails. http://ibv.vdma.org/documents/256550/5908095/2014-11-04_2_1230_Pallas+Ludens.pdf/b60f4ba6-abdf-4b05-b5ed-38756c7c8182
3. Becker, J.: Toward Fully Automated Driving: Proceedings: Toward Fully Automated Driving (2014)
4. Coelingh, E.: Intelligent Vehicles: Invited Talk (07 July 2016). https://www.youtube.com/watch?v=93lcfXrccJw
5. Darms, M.: Data fusion of environment-perception sensors for ADAS. In: Winner, H., Hakuli, S., Lotz, F., Singer, C. (eds.) Handbook of Driver Assistance Systems: Basic Information, Components and Systems for Active Safety and Comfort. Springer International Publishing, Cham (2016)
6. Dietmayer, K., Reuter, S., Nuss, D.: Representation of fused environment data. In: Winner, H., Hakuli, S., Lotz, F., Singer, C. (eds.) Handbook of Driver Assistance Systems: Basic Information, Components and Systems for Active Safety and Comfort. Springer International Publishing, Cham (2016)
7. Donges, E.: Driver behavior models. In: Winner, H., Hakuli, S., Lotz, F., Singer, C. (eds.) Handbook of Driver Assistance Systems: Basic Information, Components and Systems for Active Safety and Comfort, pp. 19–33. Springer International Publishing, Cham (2016). https://doi.org/10.1007/978-3-319-12352-3_2, https://doi.org/10.1007/978-3-662-48847-8_25
8. Gasser, T.M.: Fundamental and special legal questions for autonomous vehicles. In: Maurer, M., Gerdes, C.J., Lenz, B., Winner, H. (eds.) Autonomous Driving: Technical, Legal and Social Aspects, pp. 523–551. Springer, Berlin (2016). https://doi.org/10.1007/978-3-662-48847-8_25
9. Gehrig, S., Franke, U.: Stereovision for ADAS. In: Winner, H., Hakuli, S., Lotz, F., Singer, C. (eds.) Handbook of Driver Assistance Systems: Basic Information, Components and Systems for Active Safety and Comfort. Springer International Publishing, Cham (2016)
10. Google Inc.: Google ReCaptcha. https://www.google.com/recaptcha/intro/index.html
11. Gotzig, H., Geduld, G.: Automotive LIDAR. In: Winner, H., Hakuli, S., Lotz, F., Singer, C. (eds.) Handbook of Driver Assistance Systems: Basic Information, Components and Systems for Active Safety and Comfort. Springer International Publishing, Cham (2016)

12. Johansson, R., Nilsson, J.: The need for an environment perception block to address all ASIL levels simultaneously. In: 2016 IEEE Intelligent Vehicles Symposium (IV), pp. 1–4 (2016)
13. Lefèvre, S., Vasquez, D., Laugier, C.: A survey on motion prediction and risk assessment for intelligent vehicles. Robomech J. **1**(1), 1 (2014)
14. Noll, M., Rapps, P.: Ultrasonic sensors for a K44DAS. In: Winner, H., Hakuli, S., Lotz, F., Singer, C. (eds.) Handbook of Driver Assistance Systems: Basic Information, Components and Systems for Active Safety and Comfort. Springer International Publishing, Cham (2016)
15. Reichart, G.: Menschliche Zuverlässigkeit beim Führen von Kraftfahrzeugen Günter Reichart: Zugl.: München, Techn. Univ., Diss., 2000, *Fortschritt-Berichte*, vol. Nr. 7. VDI-Verl., Düsseldorf (2001)
16. SAE International: Taxonomy and Definitions for Terms related to On-Road Motor Vehicle Automated Driving Systems (2014)
17. Schakel, W., Knoop, V., van Arem, B.: Integrated lane change model with relaxation and synchronization. Transp. Res. Rec.: J. Transp. Res. Board **2316**, 47–57 (2012)
18. Simonite, T.: Tesla tests self-driving functions with secret updates to its customers' cars: the internet connection built into every Tesla gives the company a unique advantage in the race to develop autonomous vehicles (2016). https://www.technologyreview.com/s/601567/tesla-tests-self-driving-functions-with-secret-updates-to-its-customers-cars/
19. Statistisches Bundesamt: Verkehrsunfälle - Fachserie 8 Reihe 7 - 2015
20. Ulbrich, S., Menzel, T., Reschka, A., Schuldt, F., Maurer, M.: Defining and Substantiating the Terms Scene, Situation, and Scenario for Automated Driving. In: 2015 IEEE 18th International Conference on Intelligent Transportation Systems - (ITSC 2015), pp. 982–988 (2015). https://doi.org/10.1109/ITSC.2015.164
21. VIRES Simulationstechnologie GmbH: OpenScenario. http://www.openscenario.org
22. Wachenfeld, W., Junietz, P., Winner, H., Themann, P., Pütz, A.: Safety assurance based on an objective identification of scenarios: one approach of the PEGASUS-Project. In: Automated Vehicles Symposium. San Francisco (2016)
23. Wachenfeld, W., Winner, H.: Virtual Assessment of Automation in Field Operation: A New Runtime Validation Method. In: 10. Workshop Fahrerassistenzsysteme, p. 161 (2015)
24. Wachenfeld, W., Winner, H.: The Release of autonomous vehicles. In: Maurer M., Gerdes C.J., Lenz B., Winner H. (eds.) Autonomous Driving: Technical, Legal and Social Aspects, pp. 425–449. Springer, Berlin (2016). https://doi.org/10.1007/978-3-662-48847-8_21
25. Winner, H.: ADAS, Quo Vadis? In: Winner H., Hakuli S., Lotz F., Singer C. (eds.) Handbook of Driver Assistance Systems: Basic Information, Components and Systems for Active Safety and Comfort, pp. 1–22. Springer International Publishing, Cham (2014). https://doi.org/10.1007/978-3-319-09840-1_62-1
26. Winner, H.: Fundamentals of collision protection systems. In: Winner H., Hakuli S., Lotz F., Singer C. (eds.) Handbook of Driver Assistance Systems: Basic Information, Components and Systems for Active Safety and Comfort, pp. 1–22. Springer International Publishing, Cham (2014). https://doi.org/10.1007/978-3-319-09840-1_47-1
27. Winner, H.: Automotive RADAR. In: Winner, H., Hakuli, S., Lotz, F., Singer, C. (eds.) Handbook of Driver Assistance Systems: Basic Information, Components and Systems for Active Safety and Comfort. Springer International Publishing, Cham (2016)
28. Winner, H.: (How) Can Safety of Automated Driving be Validated? (2016). http://www.fzd.tu-darmstadt.de/media/fachgebiet_fzd/publikationen_3/2016_5/2016_Wi_Wf_Ju_ViV-Symposium_Graz.pdf
29. Winner, H.: How to address the approval trap for autonomous vehicles. In: A survey of the challenge on safety validation and releasing the autonomous vehicle (September 2015)
30. Winner, H., Geyer, S., Sefati, M.: Maße für den Sicherheitsgewinn von Fahrerassistenzsystemen. Maßstäbe des sicheren Fahrens **6** (2013)
31. Winner, H., Wachenfeld, W.: Absicherung automatischen Fahrens. In: 6. Tagung Fahrerassistenz (2013)
32. Zizer, A.: Systematische Bewertung und prototypische Nutzung von für die automatische Fahrt nicht zugänglichen Informationen für die Bewertung der automatische Fahrt

Chapter 4
Statistical Model Checking for Scenario-Based Verification of ADAS

Sebastian Gerwinn, Eike Möhlmann and Anja Sieper

Abstract The increasing complexity of advanced driver assistant systems requires a thorough investigation of a high-dimensional input space. To alleviate this problem, simulation-based approaches have recently been proposed. An overall safety assessment, however, requires the simulation results to relate to the safety of such systems in real-world scenes. To this end, we propose a rigorous method of specifying requirements for these systems depending on the environmental situation and generating statistical evidence for the safety of the system in the specified environmental situations. We demonstrate this process in an exemplary highway scenario involving decision and perception uncertainty.

4.1 Introduction

Modern automotive systems support the driver in many new ways. Assistant systems range from highlighting certain critical aspects of the current scene [1] to completely

---

This work has been conducted within the ENABLE-S3 project that has received funding from the Ecsel Joint Undertaking under grant agreement no. 692455. This joint undertaking receives support from the European Union's Horizon 2020 Research and Innovation Programme and Austria, Denmark, Germany, Finland, Czech Republic, Italy, Spain, Portugal, Poland, Ireland, Belgium, France, Netherlands, United Kingdom, Slovakia and Norway.

S. Gerwinn (✉) · E. Möhlmann · A. Sieper
OFFIS - Institute for Information Technology,
Escherweg 2, 26121 Oldenburg, Germany
e-mail: sebastian.gerwinn@offis.de

E. Möhlmann
e-mail: eike.moehlmann@offis.de

A. Sieper
e-mail: anja.sieper@offis.de

© Springer International Publishing AG, part of Springer Nature 2019
H. Waschl et al. (eds.), *Control Strategies for Advanced Driver Assistance Systems and Autonomous Driving Functions*, Lecture Notes in Control and Information Sciences 476, https://doi.org/10.1007/978-3-319-91569-2_4

taking over the actual driving task. Such features require an accurate sensing and interpretation of the surrounding environment including object detection and prediction. To achieve this task, most systems heavily rely on sensors. Consequently, the set of potential inputs to the system is as large as there are environmental situations (including their different possible appearances). As some components of the system need to interpret the sensory input, validating requirements for such components involve comparing the generated interpretation to the ground truth as provided by the environment.

However, due to the high-dimensional dependence on the environment, verifying these requirements is a major challenge in the process of rolling out these systems. Although formal methods have been investigated for addressing this challenge [14], these methods suffer from the complexity of the verification task. Also, the availability of formal models for parts of the system and—even if available—possible deviations between the model and the deployed system are additional drawbacks. To guarantee that the system satisfies the requirements, we here focus on testing by means of simulation as a possible solution. In particular, we are aiming at providing a quantitative measure of the performed simulations, thereby addressing the challenge of defining suitable metrics, see [18]. Due to the huge combination of possible inputs (or sequences thereof) as well as the high safety targets (e.g. halve the amount of accidents per year), which we would like the system to meet, the necessary amount of tests in terms of test kilometres has been estimated to be in the order of hundreds of millions of kilometres (see [15, 20]). Performing these tests is expensive and current research projects focus on an alternative employing virtual testing in terms of simulations [9, 16]. When performing tests, either virtual or real-life tests, a rigorous evaluation is necessary to provide sufficient evidence that the high safety targets are met by the system under test. That is, one has to show that the system under tests will operate safely across all possible environmental situations including all potential behaviours of other traffic participants.

In this paper, we are aiming at a two-stage approach. Building upon previous work [3, 12], the environmental situations as well as the desired (and undesired) behaviour of the system are characterized in a first step by a visual specification together with additional requirements. Although the existing visual characterization is also associated with a formal semantics, we identify several aspects, for which the characterization needs to be extended in order to enable capturing more realistic systems. In a second verification step, information about specified environmental situations as well as the system under test (SUT) is used for a particularly designed simulation of the combined system (environment+SUT) to statistically show that the system operates safely in all specified situations. More precisely, this second step performs statistical model checking of the combined system [8, 21], thereby providing a confidence statement about the satisfaction of the requirements. Importantly, both the environment and the system under test are allowed to contain abstract characterizations, thereby providing such safety statements for a large set of possible implementations (of the system) and specific situations entailed by the abstract description. Although the method in [8] can handle these under-specified systems, in this paper, we extend the evaluation method, which assumes less knowledge about the system and is therefore easily applicable in a black box setup, in which some parts of the system are only given in terms of available simulator components.

The contributions of this paper are therefore twofold. First, we identify necessary extensions of existing visual requirements formalisms in order to be applicable in more realistic settings. Second, we present an adaption of a statistical model checking technique, which can handle black box settings. Note that, although the sketched extension of the visual logic could be automatically translated into the formal description (as needed by the statistical model checking engine), we do not pursue this automatic generation within this paper.

The outline of the paper is as follows: In Sect. 4.2, we present an exemplary scenario consisting of a simple overtaking manoeuvre. We present previous work on scenario description and statistical model checking in Sect. 4.3. In Sect. 4.4, we identify necessary extensions of these ingredients for an overall safety assessment framework which allows to (1) visually describe scenarios, (2) specify safety requirements for the ego vehicle and (3) automatically assess satisfaction of the requirements via simulations. To illustrate the proposed framework, we manually translated the visual description of the scenario into a formula suitable for statistical model checking. This allows us to compute the worst-case probability of being safe within this scenario.

4.2 Running Example

To illustrate the formalization mechanism as well as the verification scheme, we use a simple overtaking example. We kept this example intentionally simple. Firstly, this facilitates illustrating the different ingredients. Secondly, at early stages of the development, available models are naturally less complex.

Within this simple example, a highly automated system is faced with the choice of either braking in front of a slower vehicle, or performing an overtaking manoeuvre (see also Fig. 4.1). More specifically, we assume that this decision is the same for all situations which are characterized by the scenario illustrated in Fig. 4.1, i.e. all distances d_b, d_f between the vehicles. Besides the lane change, the overtaking manoeuvre could also be associated a potential acceleration change in order to adapt to the velocity of the vehicle in front. Initially, however, we are interested in the safety associated purely with the lane-change manoeuvre, i.e. without a change in acceleration.

Within this example, the safety of this supervisory control depends on uncertainties, which we would like to capture for the requirements. The uncertainties considered in this example are:

- Behaviour of the vehicle behind the ego vehicle:
  – A reaction time of the vehicle behind the ego vehicle, after which it will start to decelerate, i.e. adapt to the new situation;

Fig. 4.1 Illustration of the simple overtaking scenario

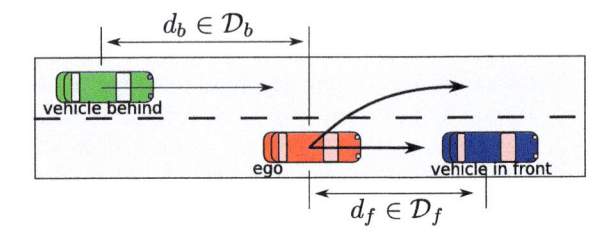

- The possibility that the green vehicle will not decelerate at all;
- The degree of deceleration the vehicle behind the ego car will perform;
• The inaccuracy of the distance as perceived by the ego vehicle.

In this example, we would like to formalize the set of situations which fall into this characterization. Provided a formalization of the dynamics of the overall scenario, we would then also like to quantify the worst-case level of safety associated with the decision for overtaking, in all these scenarios.

4.3 Related Work

As the proposed approach in Sect. 4.4 depends on previous work in the area of visual specification of operational situations and statistical model checking, we provide a brief review of these techniques in this section.

4.3.1 Visual Logic

As mentioned in the introduction, modern driver assistant systems heavily rely on sensory data, as they are required to perceive and subsequently interpret the environment in order to decide for a safe yet cost-efficient manoeuvre option.[1] For designing and also for the verification of such systems, it is important to precisely describe all possible situations a system can potentially be faced. Finding sufficiently precise abstractions is the major challenge in this context. To obtain a manageable set of scenarios, we want each scenario to encapsulate sufficiently many fine-grained behaviours. Additionally, as manually constructing a mathematical characterization—in, for example, linear time logic (LTL) or metric time logic (MTL)—of behaviours subsumed within a scenario is typically error-prone, an intu-

---

[1] In fact, the decision component has to solve a multi-criteria trade-off between travel time (speed), comfort and potentially further criteria. In this paper, however, we focus on the evaluation with respect to a safety criterion.

itive way to elicit such requirements is important to enable engineers capturing their intended requirements.

To obtain formal requirements, often logical representation is used. For the specific case of traffic situations, there exist already particularly designed frameworks in the form of multi-lane spatial logic [13] and visual logic [12]. While within the multi-lane spatial logic, one can formulate and automatically prove the safety of an abstract manoeuvre-deciding automaton; it assumes perfect knowledge of the environment of the ego vehicle and instantaneous communication between different vehicles. Visual logic on the other hand combines a visual representation of assumptions on the static and dynamic aspects of possible environment scenarios with the explicit specification of the communication between different actors in the scene.

In the *Visual Logic*, specifications are composed of a sequence of atoms, corresponding to the different scenes of a scenario. Each such *atom* consists of a traffic view (TV) and a communication description. With these atoms, it is possible to specify a traffic scenario. The *traffic view* comprises a visualization and a formal representation, capturing the geometric relations between the different objects. The *communication description* represents the communication between the different involved communication partners by using the well-established formalism of live sequence charts (LSCs) [3]. In contrast to using a single illustration as in Fig. 4.1, visual logic provides an abstract description of the scenario by stating sequences of these atoms.

In Fig. 4.2, the traffic views of a successful overtaking manoeuvre are shown. The initial situation is stated in Fig. 4.2a—here, the ego vehicle is coloured red. Figure 4.2b describes the situation where the ego vehicle overlaps the lane separator during the manoeuvre and therefore enters the other lane. In the last key scene, the ego car is next to the blue car indicating a successful overtaking (cf. Fig. 4.2c). In the last traffic view, the ego vehicle should be in front of the blue vehicle on the same

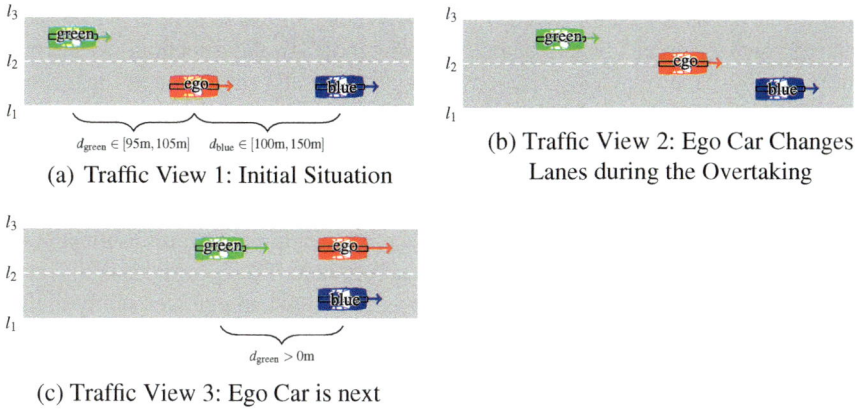

(a) Traffic View 1: Initial Situation

(b) Traffic View 2: Ego Car Changes Lanes during the Overtaking

(c) Traffic View 3: Ego Car is next to blue Car during the Overtaking

Fig. 4.2 Traffic views of the highway scenario in case of successful overtaking manoeuvre

Table 4.1 Formalization of the successful overtaking manoeuvre (cf. Fig. 4.2)

Traffic view	Longitudinal positions	Lateral positions	Positions between lane separators and vehicles
TV 1	ego $after_{\leftrightarrow}$ green ego $before_{\leftrightarrow}$ blue	ego $before_{\updownarrow}$ green ego $equals_{\updownarrow}$ blue	green $is\ between\ l_2$ and l_3 ego $is\ between\ l_1$ and l_2 blue $is\ between\ l_1$ and l_2
TV 2	no changes, see TV 1	ego $meets_{\updownarrow}$ green ego $meets_{\updownarrow}$ blue	green: no change ego $is\ on\ l_2$ blue: no change
TV 3	ego $after_{\leftrightarrow}$ green ego $equals_{\leftrightarrow}$ blue	ego $equals_{\updownarrow}$ green ego $after_{\updownarrow}$ blue	green: no change ego $is\ between\ l_2$ and l_3 blue : no change

lane with at least the same speed. Although, the whole overtaking manoeuvre could consist of further key scenes, for our example it suffices to end with the scene depicted in Fig. 4.2c. All traffic views graphically represent relative positions between the ego and other vehicles.

Relative positions as well as additional constraints between lane separators and vehicles can be formalized in forms of either abstract relations or absolute distances (for the technical details we refer to [12]). Absolute distances can be annotated within the visualization (cf. Fig. 4.2). For the successful overtaking manoeuvre the abstract relations are listed in Table 4.1. For the initial situation, we will explain the formalization in detail. In the figure, the ego vehicle is in front of the green vehicle (longitudinal direction), which corresponds to the predicate ego $after_{\leftrightarrow}$ green, where the arrow indicates the direction (i.e. longitudinal) of the relation and the measurement takes the minimal distance of the bounding boxes. Furthermore, the ego vehicle is longitudinal behind the blue vehicle (ego $before_{\leftrightarrow}$ blue). In lateral direction, the ego vehicle is at the same position as the blue one (ego $equals_{\updownarrow}$ blue) and right of the green one (ego $before_{\updownarrow}$ green), where again the arrow indicates the direction, i.e. lateral. The positions of the vehicles in relation to the lane separators are given by the formulas green $is\ between\ l_2$ and l_3, ego $is\ between\ l_1$ and l_2 and blue $is\ between\ l_1$ and l_2. Figures 4.3 and 4.4 show the key scenes for a frontal and a side-by-side collision,[2] respectively. They can be formalized in the same way as the key scenes for the successful manoeuvre. Due to the lack of communication description in our scenario, we do not have any LSCs. Using the semantics of the visual logic, pictures as illustrated in Fig. 4.2 can be translated directly into logical formulae. These formulae in turn can be checked for satisfiability, i.e. whether the assumed system operates safely in the characterized environment. As the expressiveness of the logic is limited, we briefly review a logical framework capable of specifying additional aspects such as richer dynamics and degrees of freedom (probabilistic and non-deterministic).

---

[2]To simplify the example, we denote by collision that the occupied regions of the cars overlap.

4 Statistical Model Checking for Scenario-Based Verification of ADAS 73

(a) Traffic View 1: Initial Situation

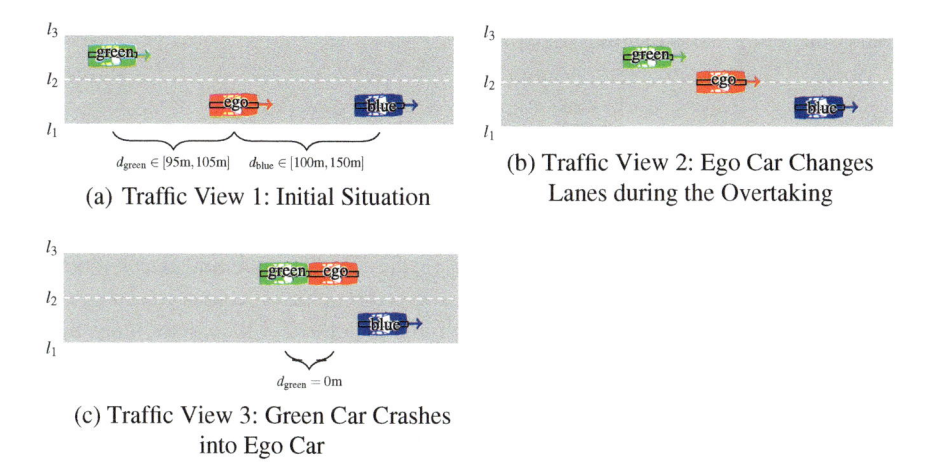

(b) Traffic View 2: Ego Car Changes Lanes during the Overtaking

(c) Traffic View 3: Green Car Crashes into Ego Car

Fig. 4.3 Traffic views of the highway scenario in case of a rear-end collision

(a) Traffic View 1: Initial Situation

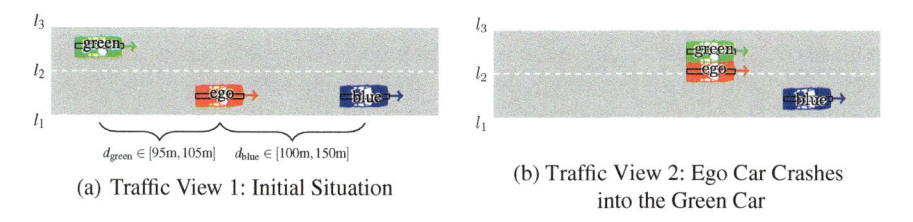

(b) Traffic View 2: Ego Car Crashes into the Green Car

Fig. 4.4 Traffic views of the highway scenario in case of a side-by-side collision

4.3.2 Formal Characterization Using Stochastic Satisfiability Modulo Theory (SSMT)

Using the abstract representation and the semantics of the visual logic, one can already specify a large set of possible behaviours of different traffic participants. Nevertheless, some aspects of the uncertainties listed in Sect. 4.2 cannot be represented in the current version of the visual logic. In the area of formal verification of (stochastic) hybrid systems, the formalism of stochastic satisfiability modulo theory (SSMT) [10] has been used to describe the dynamics as a result of the interactions between the environment and the system under test, where discrete controllers act within a continuous, potentially stochastic environment and therefore offers a sufficiently rich language to describe the necessary artefacts. Although other formalisms offer a similar degree of flexibility (e.g. UPPAAL stratego [7]), we focus on SSMT for now.

Indeed, using SSMT we can specify requirements involving dynamics using differential equations, switching between different dynamics, and setting new variable values using either stochastic transitions and valuations or completely non-deterministic versions. We illustrate this modelling by means of an example, which we have already used in a similar setting in [8]:

$$\forall_{d_b \in \mathcal{D}_b} \forall_{d_f \in \mathcal{D}_f} ⨄_{\Delta t \sim \beta(\alpha_t, \beta_t)} ⨄_{a_b \sim \beta(\alpha_a, \beta_a)} \forall_{\varepsilon_p \in \Delta_p} \forall_{\varepsilon_v \in \Delta_v} :$$
$$c(d_b, d_f, \Delta_t, a_b, \varepsilon_p, \varepsilon_v) < \theta \quad (4.1)$$

Here, d_b, d_f denote the distance to the vehicles behind and in front of the ego vehicle, ε_p, ε_v denote the inaccuracy of the perception of the position and velocity, and Δ_t, a_b denote random variables characterizing the reaction time and deceleration of the vehicle behind the ego vehicle in case the ego vehicle decides for an overtaking manoeuvre. Additionally, a cost function c is defined, which determines the costs associated with a single evolution of the dynamics. For instance, c can quantify the severity of a collision in the trace (or evaluating to 0 in case no collision has occurred). Alternatively, a safety specification can directly be integrated into the cost function by comparing the severity to the allowed target (in the above formula θ). The non-deterministic choices of the variables d_b, d_f, ε_p, ε_v need to be resolved pessimistically. That is, we would like the property to hold for all possible choices of the values for these variables, indicated by the universal quantifier $\forall$. Other variables (in the above equation: Δ_t, a_b) are modelled as probabilistic choices. For these variables, we would like the property to hold *on average*, indicated by the randomized quantifier $⨄$. For these variables, we have to specify the probability distribution, from which the values are chosen randomly. In the example, these are chosen according to a beta distribution $\sim \beta(\alpha_t, \beta_t)$, $\sim \beta(\alpha_a, \beta_a)$. Finally, one specifies a quantifier-free formula ($c(d_b, d_f, \Delta_t, a_b, \varepsilon_p, \varepsilon_v) < \theta$ in Eq. 4.1), which evaluates a particular choice of variables and encapsulates the detailed dynamics. Using these ingredients, we are able to specify desired properties. For example, that the criticality (measured by the criticality cost function c) of all possible realizations of a scenario should be below a given threshold (θ)—at least on average. A formula as in Eq. 4.1 is then associated with a value, indicating the probability of satisfaction.[3] Instead of representing the statements in terms of universally, existentially and randomized quantifiers, we could also represent the probability of satisfaction as a nested, noisy optimization problem. That is, Eq. 4.1 can be equivalently be written as:

$$\min_{d_b \in \mathcal{D}_b} \min_{d_f \in \mathcal{D}_f} \mathbb{E}_{\Delta t \sim \beta(\alpha_t, \beta_t)} \left[\mathbb{E}_{a_b \sim \beta(\alpha_a, \beta_a)} \left[\min_{\varepsilon_p \in \Delta_p} \min_{\varepsilon_v \in \Delta_v} \mathbb{1}_{c(d_b, d_f, \Delta_t, a_b, \varepsilon_p, \varepsilon_v) < \theta}(d_b, d_f, \Delta_t, a_b, \varepsilon_p, \varepsilon_v) \right] \right] \quad (4.2)$$

Here, min refers to the pessimistic resolution of the non-deterministic choices, such as distances, whereas $\mathbb{E}_{X \sim \beta(\alpha, \gamma)}$ denotes the expected value across a random variable X which is distributed according to a β distribution with parameters α and γ. $\mathbb{1}_{c(x) < \theta}(x)$ denotes the indicator function which evaluates to one for all settings x

---

[3] Note that there is a difference between 'for all on average' and 'on average for all' or worst-average-case vs average-worst-case. The 'average-worst-case' is generally a more strict requirement, i.e. has a lower probability to be satisfied compared to the 'worst-average-case'.

for which the cost function c is smaller than a given threshold θ and zero otherwise. For a more formal treatment of the semantics, we refer to [8].

4.3.3 SSMT Solving Using Statistical Model Checking

Due to the heterogeneity of dynamic effects and the tight interaction of variables of different types, the analysis of stochastic hybrid systems is notoriously difficult, calling for automated verification methods. With the SSMT formalization and the corresponding constraint solvers, it is possible to model networks of hybrid automata which include both discrete probabilistic branching and non-deterministic decisions on transitions. The computational complexity of discharging the proof obligations by exact and exhaustive stochastic constraint solving, however, often is prohibitive. In [8], we extended the model checking procedure for solving SSMT problems to a more scalable version based on simulations and a corresponding statistical analysis, i.e. statistical model checking (SMC) [21]. In contrast to classical model checking, the SMC method is based on samples using an underlying simulator and therefore generates results which can be guaranteed with a certain level of confidence only, i.e. have a residual probability of being false.

As a proof that a formula such as Eq. 4.1 holds is notoriously difficult, existing solutions [10, 17] typically do not scale well to more complex systems involving multiple randomized and other quantified variables. When solving such a formula using simulations, the resulting statement is only of statistical nature. To illustrate this, let p be the exact value of the formula in question (e.g. Eq. 4.1). Using statistical model checking, we obtain a confidence interval $[\underline{p}_\delta, \overline{p}_\delta]$ which contains the true value only with the specified confidence $1 - \delta$. More precisely, the probability of generating a set of samples such that the calculated confidence interval, based on this set of samples, will not contain the true value can be bounded by δ. However, in an SSMT formula (see Eq. 4.1), we allow not only purely random influences, but also pure non-deterministic influences which have to be resolved either optimistically or pessimistically as reflected by the maximum and minimum operator in Eq. 4.2. In such a setting, pure sampling is not directly applicable. Hence, in [8], we incorporated methods from noisy optimization, specifically algorithms such as the upper confidence bound strategy for multi-armed bandits (see [2]). In this extension [8], we iteratively built up a forest of decision trees, each of which represents a variable within the quantifier prefix ($d_b, d_f, \Delta t, a_b, \varepsilon_p \varepsilon_v$ in Eq. 4.1). Each node contains a confidence interval, valid for the given prefix in the forest. The upper confidence strategy then decides in each iteration, i.e. a newly generated sample, for the path with the most promising interval (upper interval bound for the maximum and lower bound for the minimum) for further inspection. We refer to [8] for the technical details. However, it is worth mentioning that this method requires detailed knowledge about the probability distributions associated with the randomized quantifier as well as about the underlying dynamics. Therefore, we refer to this simulation-based evaluation strategy as white box model checking. However, in a virtual simulation-based environment, parts of the simulation are only available via black box simulation. These simulations might still contain randomness, for example, the behaviour of other vehicles within a scenario might be subject to a particularly designed probabilistic dynamic model which is only available via a black box simulation.

4.4 Methods

As mentioned, to describe the requirements necessary for a highly automated function, for example, the supervisory control deciding for an overtaking manoeuvre (see Sect. 4.2), we are aiming at formal representation of the possible scenario trajectories followed by a simulation-based evaluation of these requirements. Building upon previous work, we first need additional annotations for the visual logic to capture possible uncertainties as listed in Sect. 4.2. Also, to support a more black box setting, we need to extend the statistical model checking scheme.

4.4.1 Traffic Sequence Charts

As our main goal is to employ statistical model checking for scenario-based verification, we need to somehow construct the input (an SSMT formula) suitable for the model checking. This can be done either (1) manually, by an expert who—with a certain traffic scenario in mind—encodes all aspects including different settings, objects, dynamics and requirements into a large formula or (2) (semi-)automatically, from a visually described traffic scenario with formal semantics. We favour (2) over (1) because it is more practical since (1) requires the expert to have both, intimate knowledge about SSMT as well as the domain and is more error-prone due to the sheer amount of aspects. To support (2), we have identified the need for

(**N1**) a visual representation of a traffic scenario (geometric configuration, etc.),
(**N2**) a possibility to represent feasible and intended behaviours,
(**N3**) a possibility to represent either (abstract) dynamics or (concrete) dynamical models of all objects—including the ego vehicle—within the traffic scenarios, and
(**N4**) a possibility to represent requirements that have to be fulfilled —for example, by the ego vehicle.

The visual logic is capable of N1 and partially of N2 by building sequences or even trees of atoms mixed with live sequence charts. However, it does support neither N3 nor N4. Hybrid automata [11] could serve as a preliminary means for N3. Nevertheless, we propose the use of a visual tool for capturing N4 more intuitively. All these needs have to be served by a larger formalism that we call *traffic sequence charts (TSCs)*.[4] In Fig. 4.1, we have already visualized the atoms for a successful overtak-

---

[4] For recent development on the topic of traffic sequence charts, we refer to [4–6].

4 Statistical Model Checking for Scenario-Based Verification of ADAS

ing situation in form of the visual logic. For TSCs, we have to represent additional information that is summarized in Table 4.2 which states for every parameter of the scenario the quantification, the domain and the distribution—if available. Figure 4.5 visualizes the successful overtaking extended with annotation needed for the model checking. On the other hand, Fig. 4.6 visualizes that a collision occurred during the overtaking. In the setting of the overtaking manoeuvre, the annotations state the relative positions, velocities and accelerations of the involved vehicles. Additionally, the green vehicle is annotated with the probability with which it might not react to the ego vehicle when entering the same lane. For illustrative purpose, we call this *rowdy*.

Table 4.2 Summary of the quantifier, domains and distributions

Parameter	Unit	Quantifier	Domain	Distribution
x_{rowdy}	–	∃	[0, 1]	$\mathcal{U}[\cdot]$
Δ_t	s	∃	[0.1, 1.201]	$\beta(2, 4)$
a_b	m/s^2	∃	[−4.3, −3.1]	$\beta(3, 3)$
d_b	m	∀	[−105, −95]	–
d_f	m	∀	[100, 150]	–
ε_b	m	∀	[−2, 2]	–
ε_f	m	∀	[−2, 2]	–

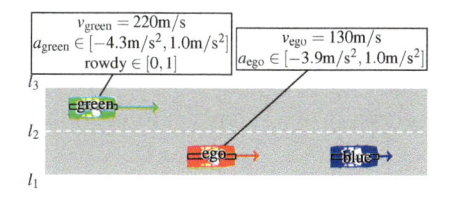

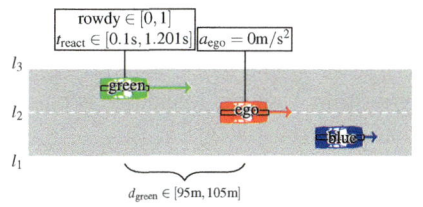

(a) Traffic View 1: Initial Situation

(b) Traffic View 2: Ego Car Changes Lanes during the Overtaking

(c) Traffic View 3: Ego Car Changed Lanes and Green Car Starts to React

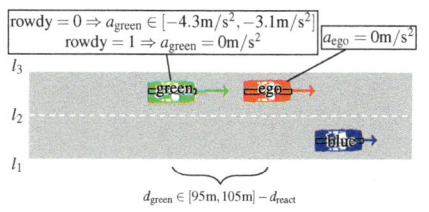

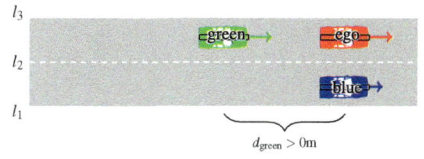

(d) Traffic View 4: Ego Car and Green Car did not Crash

Fig. 4.5 Traffic views of the highway scenario in case of successful overtaking

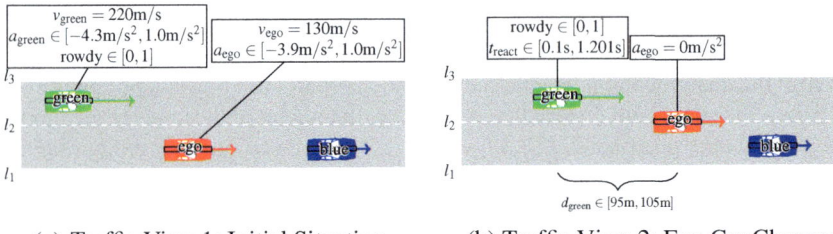

(a) Traffic View 1: Initial Situation

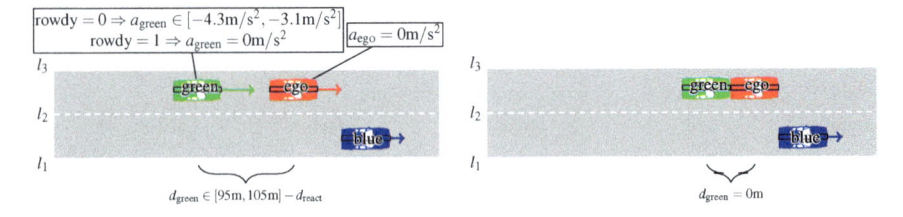

(b) Traffic View 2: Ego Car Changes Lanes during the Overtaking

(c) Traffic View 3: Ego Car Changed Lanes and Green Car Starts to React

(d) Traffic View 4: Green Car Crashes into Ego Car

Fig. 4.6 Traffic views of the highway scenario in case of rear-end collision

Together with a—possibly abstract, non-determinism introducing—dynamical model of the vehicles, we have all information at hand to generate a formal representation for the method presented in the next section. This formal representation is a means to answer the following question.

Question: Can we bound the worst-case probability of a severe collision given a fixed control strategy?

That is, we are interested in finding probability of a severe collision (as measured by the cost function, see Eq. 4.1). This probability is the result of averaging across the probabilistic choices (such as reaction time and accelerations). As this probability additionally depends on non-deterministic choices, we are interested in determining the worst case across all these free variables (such as distances and sensor inaccuracies).

4.4.2 Statistical Model Checking

In Sect. 4.3.3, we described a method how to statistically verify a system to satisfy properties. This satisfaction problem can be specified using stochastic satisfiability modulo theory (SSMT). For scalability reasons as well as exploiting available simulators, the presented method makes extensive use of a statistical analysis on generated samples. However, to determine the probability of satisfaction, one requires detailed knowledge about the system at hand. In particular, it requires bounding the effect of choosing an interval as input (either as a parameter or as a set of states). Especially when models for vehicle or controller dynamics are involved, such computation would require a white box setting, in which the details of the dynamics are known. In practice, however, such knowledge might not be available, as corresponding simulators are only available as a black box.

Therefore, in this section, we present a different evaluation strategy for problems which are also specified in terms of SSMT formulas such as the one used in the evaluation (see Eq. 4.6), without assuming to be able to bound the effect of intervals. Similarly, as mentioned, the probability distributions might only implicitly be given in terms of sample generators, i.e. simulation models. However, in order to compute meaningful uncertainty intervals, we assume that the cost function, which evaluates individual simulation runs, can still be bounded. For example, when testing a binary property such as collision occurrence, bounds are directly given as the cost function can only produce two outcomes: 1 corresponding to a crash and 0 for a crash-free simulation run.

Although we only assume having black box simulations, we are still interested in worst-case statements over sets of environmental situations or sensor inaccuracies as illustrated in Sect. 4.4.1. That is, we require a sampling mechanism to resolve these non-deterministic choices either pessimistically or optimistically. To this end, we adopt the algorithm presented in [19] by Vidyasagar, which we repeat here for completeness. To illustrate this algorithm, consider the following noisy optimization problem for a given cost function ϕ and probability distribution P:

$$\max_{x \in \mathcal{X}} \int \phi(x, y) \mathrm{d} P_Y(y|x) \tag{4.3}$$

Here, the optimization problem consists of finding a variable $x \in \mathcal{X}$ for which the expected costs are maximal. The expected costs in turn can be expressed by a cost function ϕ averaged across random influences. These random influences can be described by a conditional distribution $P_Y(\cdot|x)$ which specifies how these random influences are distributed for a given choice of x.

To solve this, Vidyasagar proposed to sample x from a known distribution, e.g. a uniform distribution. For each of these samples, one computes an empirical average by sampling again from the conditional distribution $P_Y(\cdot|x)$ and then estimates the maximum by taking the maximum over the x-samples (see Algorithm 1).

It can be shown that using Algorithm 1, the following holds for the estimate $\hat{p}$:

$$P_S\left(P_X\left(\hat{p} + \varepsilon \leq \int \phi(x, y) \mathrm{d} P_Y(y|x)\right) \leq \alpha\right) \geq 1 - \delta, \tag{4.4}$$

where P_S denotes the probability as a result of the sample $S = (x_n, y_m), n = 1, \ldots, N$, $m = 1, \ldots, M$ and P_X denotes the chosen probability distribution for sampling x.

Algorithm 1 Randomized maximization algorithm for $\max_{x \in \mathcal{X}} \int \phi(x,y) \mathrm{d}P_Y(y|x)$

function RandMax(Confidence $1-\delta$, Accuracy ε, Residual domain α, Property ϕ)
 $\quad N \leftarrow \frac{\log \frac{2}{\delta}}{\log \frac{1}{1-\alpha}}, \; M \leftarrow \frac{1}{2\varepsilon^2} \log \frac{4N}{\delta}$
 for $n = 1, \ldots, N$ **do**
  Draw optimization samples x_n uniformly over the domain $\mathcal{X}$
  for $m = 1, \ldots, M$ **do**
   Draw random sample y_m from the distribution $P_Y(\cdot|x_n)$
   $\phi_{n,m} \leftarrow \phi(x_n, y_m)$ $\quad\quad\quad\quad\quad\quad\quad\quad\quad\quad$ ▷ Evaluate property
  end for
  $\hat{\phi}_n = \frac{1}{M} \sum_m \phi_{n,m}$ $\quad\quad\quad\quad\quad$ ▷ Empirical average across m samples
 end for
 return $\hat{p} = \max_n(\hat{\phi}_n)$
end function

The inner probability expresses the likelihood that there is another x value that corresponds to an expectation ($\int \phi(x,y) \mathrm{d}P_Y$) which is more than ε larger. If we choose a uniform distribution for generating the variable x over which we want to maximize, this probability relates to the volume of the set of x-values with potentially better probability of satisfaction. The outer probability in Eq. 4.4 additionally bounds the probability that this statement is only a fluke as the result of being based on finitely many samples MN. Analogously, we can obtain an algorithm and statistical guarantee for minimization problems, by replacing the maximum with the minimum in Algorithm 1 and $\hat{p} - \varepsilon$ in the guarantee. Algorithm 1 takes as input an accuracy ε. Alternatively, one can also specify the number of samples K for which sufficient computational budget is available. In this case, the accuracy ε as well as N, M can be computed as follows:

$$N = \frac{\log \frac{2}{\delta}}{\log \frac{1}{1-\alpha}}, \quad M = \frac{K}{N}, \quad \varepsilon = \sqrt{\frac{1}{2M} \log \frac{4N}{\delta}}$$

In order to use this algorithm to evaluate SSMT formulas, we use the optimization semantics; that is, a universal quantifier corresponds to a minimization problem and an existential quantifier to a maximization problem. Albeit, we are only able to handle problems of the form of possibly multiple existential or universal quantifiers followed by randomized ones. This poses no major restriction, because the type of problems we are considering here can be formulated as follows: is there a bound on the likelihood of a system satisfying a certain property that holds for all environmental scenarios? Mathematically, this corresponds to the noisy optimization problem above. Nevertheless, future work will generalize the method also to nested quantifier problems.

Comparison with White Box Statistical Model Checking

Note that the statement of Eq. 4.4 differs from the confidence interval obtained from the method presented in Sect. 4.3.3. The difference is that the former takes two uncertainty parameter α, ε in addition to confidence level δ. While the *white box* method, presented in Sect. 4.3.3, uses more detailed knowledge about the regularity of the system, the *black box* method of algorithm Algorithm 1 has no such knowledge. Therefore, the confidence statement of Eq. 4.4 is of different nature. While it is possible for the system to behave completely different for different non-deterministic choices, the chances of choosing the unlucky ones (in terms of the analysis goal) are bounded. As values of the non-deterministic variables are chosen uniformly at random, the confidence statement measures the chances of the black box method generating an unlucky sample in terms of a residual domain (α) and accuracy (ε), thereby replacing the maximum across X variables with the likelihood of generating worse samples. By replacing the maximum with a likelihood statement, strong assumptions about the regularity can be dropped at the cost of a less specific confidence statement. The white box method on the other hand directly provides a confidence statement of the following form:

$$P\left(p \in [\hat{p} - \varepsilon, \hat{p}^w + \varepsilon^w]\right) \geq 1 - \delta^w \tag{4.5}$$

where p is the true (but unknown) value of equations such as Eq. 4.3, $\hat{p}^w$ is the point estimate of the white box method, ε^w is the accuracy, and $1 - \delta^w$ is the confidence level of the white box method.

In order to still be able to compare both methods, we summarize the probability for the estimate $\hat{p}$ being more than ε off. Specifically, the probability the sample being not a fluke and simultaneously the expectation being below the epsilon corrected estimate can be bounded by $(1 - \alpha)(1 - \delta)$. This corresponds to multiplying the two probabilities in Eq. 4.4. For the comparison, we interpret this as a confidence about the actual maximum and map the confidence interval obtained by the white box method to the interval of the black box method. More specifically, we set $(1 - \alpha) := (1 - \delta^b) := \sqrt{1 - \delta^w}$, where $1 - \delta^b$ refers to the confidence of the black box method and α measures the residual domain for the black box method. Using these settings, we obtain an estimate $\hat{p} \pm \varepsilon$ of comparable confidence.

From a practical perspective, the white box model checking technique is not necessarily able to draw samples from the specified probability distributions, as it has only access to the symbolic representation of the density functions. As a result, it uses importance sampling by using a uniform distribution as a proposal and later corrects for this different sample distribution using the appropriate importance weight (see [8]). In a black box setting, the situation is reversed. Here, one has access to a sample-generating function, but not necessarily to the corresponding density. Therefore, to compare the two evaluation strategies, we artificially also used importance sampling in the black box setting. Consequently, to calculate the accuracy ε, we have to correct for the potentially larger range of values in the empirical mean resulting in a potentially larger confidence interval.

4.5 Results

So far, we have reported on (1) how to describe traffic scenarios tailored for automatic analysis and (2) how to apply statistical model checking as a means of quantifying safety of advanced driver assistant systems and autonomous driving functions. In this section, we analyse the scenario described in Sect. 4.2 focusing on the ego car's strategy to overtake. Specifically, the property we are interested in is the crash costs being lower than a certain threshold $\theta = 1e^{-2}$. Therefore, we fix the behaviour of the ego car and derive the following SSMT formula:

$$\forall_{d_b \in \mathcal{D}_b} \forall_{\varepsilon_b \in \Delta_p} \exists_{x_{\text{rowdy}} \sim \mathcal{U}[0,1]} \exists_{\Delta_t \sim \beta(\alpha_t, \beta_t)} \exists_{a_b \sim \beta(\alpha_a, \beta_a)} : \\ c_{\text{rowdy}}(d_b, \varepsilon_b, x_{\text{rowdy}}, \Delta_t, a_b) < \theta \quad (4.6)$$

which realizes the likelihood of encountering a rowdy using a (continuous) uniform distribution of the random variable x_{rowdy} being above a certain threshold θ_{rowdy}. The cost function is given by

$$c_{\text{rowdy}}(d_b, \varepsilon_b, x_{\text{rowdy}}, \Delta_t, a_b) := \begin{cases} c(d_b, \varepsilon_b, \infty, 0) & \text{if } x_{\text{rowdy}} \geq \theta_{\text{rowdy}} \\ c(d_b, \varepsilon_b, \Delta_t, a_b) & \text{if } x_{\text{rowdy}} < \theta_{\text{rowdy}} \end{cases}$$

which computes the severity of a collision depending on the particular evolution of the scenario. This in turn depends on whether the green car happens to be a rowdy (no reaction, i.e. $\Delta_t = \infty$, $a_b = 0$) or not (normal reaction) as well as the specific distances and inaccuracies chosen for this evolution.

The collision costs $c(d_b, \varepsilon_b, \Delta_t, a_b)$ can then be computed as follows: given $d_b, \varepsilon_b, \Delta_t, a_b$, we compute whether a collision occurs. This corresponds to checking whether the two cars meet at any time while we propagate the initial state d_{b0} according to the car's dynamics. In our settings, we use a simple dynamic model where the distance evolves according to the relative velocities $\dot{d}_b = v_{\text{ego}} - v_{\text{green}}$ for Δ_t time and then additionally the velocity decreases with $\dot{v}_{\text{green}} = a_b$ until $v_{\text{green}} \leq v_{\text{ego}}$. If at any time the cars collide, then the severity of the impact is given by the following cost formula

$$\texttt{costs}(v_{\text{green}}, v_{\text{ego}}) := |v_{\text{ego}} - v_{\text{green}}| \max(v_{\text{ego}}, v_{\text{green}}).$$

We have prototypically implemented the white box-based statistical model checking approach as described in Sect. 4.3.3 as well as the black box-based one described in Sect. 4.4.2. To compare these two methods, we ran the analysis for both model checking methods (WhiteBox and BlackBox) and different likelihoods of encountering a rowdy $\theta_{\text{rowdy}} \in \{0, 0.1, 0.5, 1.0\}$. The results are depicted in Fig. 4.7 and the actual quantitative estimates are summarized in Table 4.3. Note that we have included results of the black box-based method with more samples to obtain a comparable accuracy of the statement.

4 Statistical Model Checking for Scenario-Based Verification of ADAS 83

(a) Confidence Intervals wrt. Number of Samples ($\theta_{\mathrm{rowdy}} = 0.0$)

(b) Width of Confidence Intervals wrt. Number of Samples ($\theta_{\mathrm{rowdy}} = 0.0$)

(c) Confidence Intervals wrt. Number of Samples ($\theta_{\mathrm{rowdy}} = 0.1$)

(d) Width of Confidence Intervals wrt. Number of Samples ($\theta_{\mathrm{rowdy}} = 0.1$)

(e) Confidence Intervals wrt. Number of Samples ($\theta_{\mathrm{rowdy}} = 0.5$)

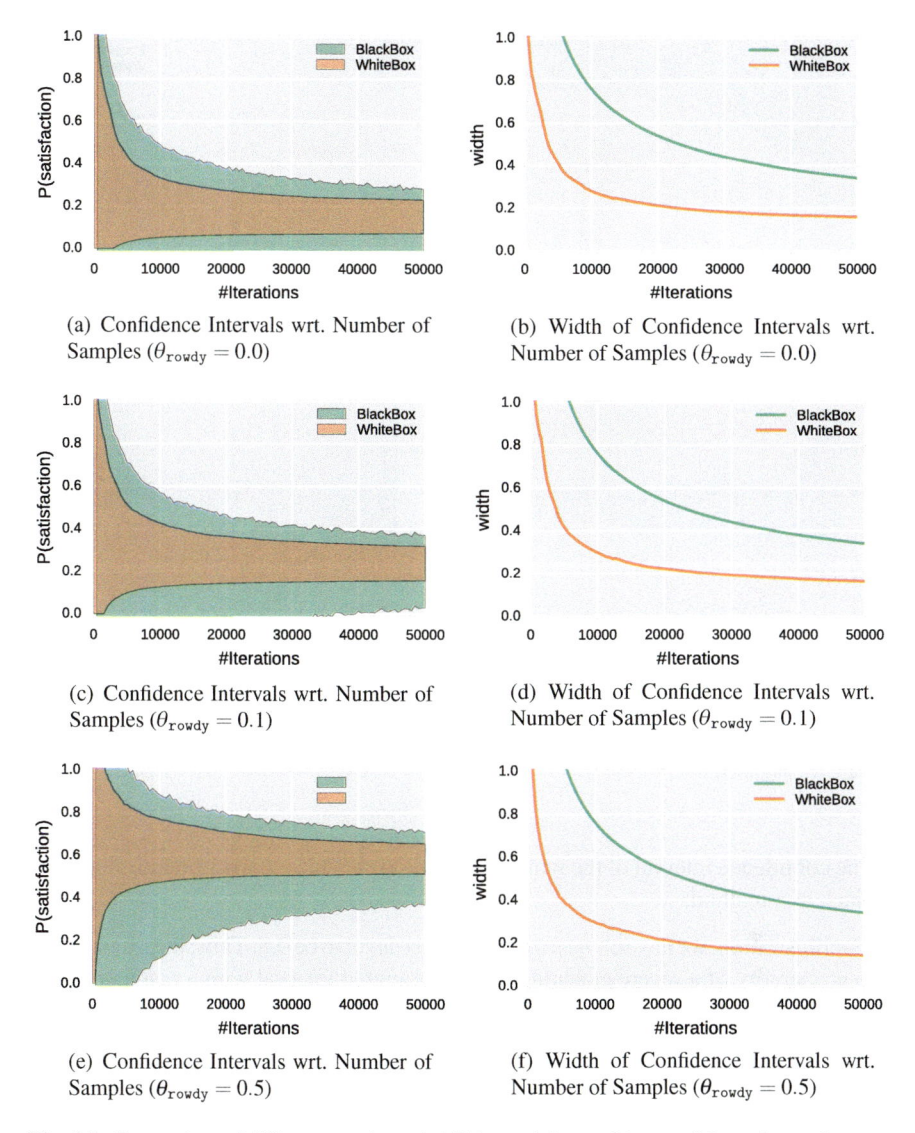

(f) Width of Confidence Intervals wrt. Number of Samples ($\theta_{\mathrm{rowdy}} = 0.5$)

Fig. 4.7 Comparison of different rowdy probabilities and the resulting confidence intervals

Using these analyses, we are therefore able to make the following statements about the hypothetical controller strategy:

- The estimated probability of a collision increases (as expected) with increasing likelihood of the green vehicle being a 'rowdy'.
- The confidence intervals produced by the white box method converge faster than for the black box method.

(g) Confidence Intervals wrt. Number of Samples ($\theta_{\text{rowdy}} = 1.0$)

(h) Width of Confidence Intervals wrt. Number of Samples ($\theta_{\text{rowdy}} = 1.0$)

(i) Confidence Intervals for the Black Box Method

(j) Confidence Intervals for the White Box Method

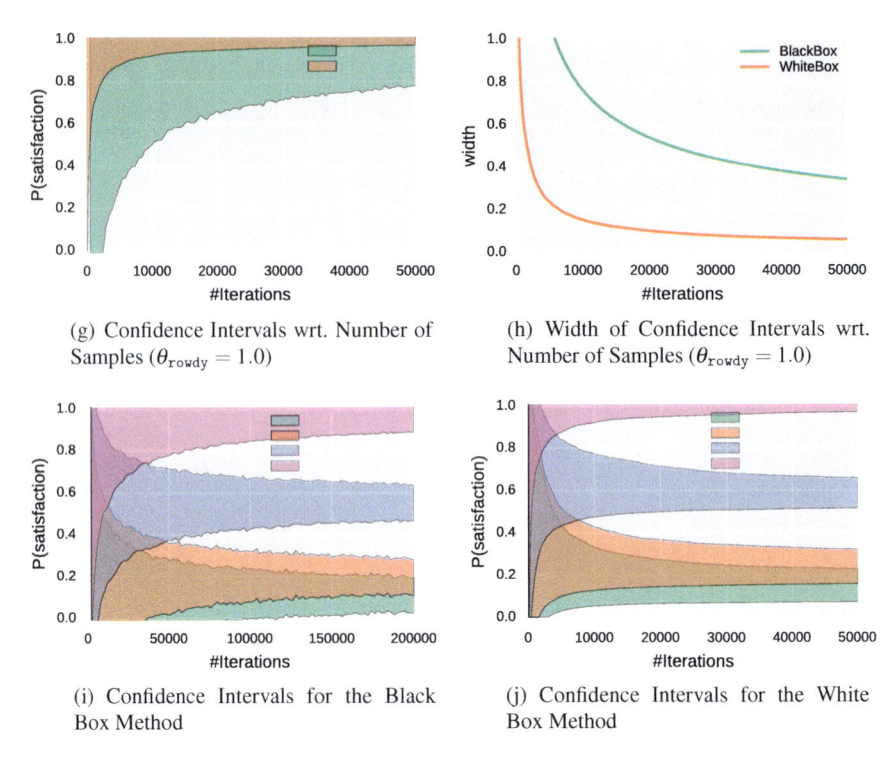

Fig. 4.7 (continued)

- The confidence interval of the white box method provides sharper bounds than the black box method.

For illustration, we have chosen the confidence level to be 0.8. Thus, the true worst-case probability of a severe crash lies in the computed interval with a confidence of at least 0.8. Or in other words, the likelihood that our confidence interval does not contain the real estimate—due to unlucky samples—can be bounded by 0.2. For practical problems, we would typically require (1) a higher confidence and, thus, (2) more samples. Nonetheless, using the data we can already conclude that deciding for an overtaking manoeuvre with the chosen acceleration in the situations characterized by the different distances (as specified in the scenario) is most likely unsatisfactory. In particular, if the vehicle behind the ego vehicle has a chance not to decelerate, the likelihood of a severe crash is increased significantly. Although, this behaviour could be anticipated beforehand, in a more realistic setting, it might not be clear, if a controller is sufficiently safe. The scenario and controller are mainly chosen to compare the different methods of evaluation on a setting having non-deterministic, probabilistic and nonlinear influences.

Table 4.3 Upper and lower bounds on the worst-case likelihood of a severe collision corresponding to the respective confidence intervals (rounded to two digits). Note that increasing threshold θ_{rowdy} actually decreases the chances of a 'rowdy' due to the uniform sampling of x_{rowdy} and decision rule $x_{\text{rowdy}} < \theta_{\text{rowdy}}$

Method	Rowdy threshold	No. samples	Upper bound	Lower bounds
WhiteBox	$\theta_{\text{rowdy}} = 0.0$	50,000	$P(Eq.\,(6)) \le 0.23$	$P(Eq.\,(6)) \ge 0.07$
WhiteBox	$\theta_{\text{rowdy}} = 0.1$	50,000	$P(Eq.\,(6)) \le 0.32$	$P(Eq.\,(6)) \ge 0.16$
WhiteBox	$\theta_{\text{rowdy}} = 0.5$	50,000	$P(Eq.\,(6)) \le 0.66$	$P(Eq.\,(6)) \ge 0.51$
WhiteBox	$\theta_{\text{rowdy}} = 1.0$	50,000	$P(Eq.\,(6)) \le 1.0$	$P(Eq.\,(6)) \ge 0.97$
BlackBox	$\theta_{\text{rowdy}} = 0.0$	50,000	$P(Eq.\,(6)) \le 0.28$	$P(Eq.\,(6)) \ge 0$
BlackBox	$\theta_{\text{rowdy}} = 0.1$	50,000	$P(Eq.\,(6)) \le 0.37$	$P(Eq.\,(6)) \ge 0.03$
BlackBox	$\theta_{\text{rowdy}} = 0.5$	50,000	$P(Eq.\,(6)) \le 0.71$	$P(Eq.\,(6)) \ge 0.37$
BlackBox	$\theta_{\text{rowdy}} = 1.0$	50,000	$P(Eq.\,(6)) \le 1$	$P(Eq.\,(6)) \ge 0.78$
BlackBox	$\theta_{\text{rowdy}} = 0.0$	200,000	$P(Eq.\,(6)) \le 0.19$	$P(Eq.\,(6)) \ge 0.02$
BlackBox	$\theta_{\text{rowdy}} = 0.1$	200,000	$P(Eq.\,(6)) \le 0.28$	$P(Eq.\,(6)) \ge 0.11$
BlackBox	$\theta_{\text{rowdy}} = 0.5$	200,000	$P(Eq.\,(6)) \le 0.63$	$P(Eq.\,(6)) \ge 0.46$
BlackBox	$\theta_{\text{rowdy}} = 1.0$	200,000	$P(Eq.\,(6)) \le 1$	$P(Eq.\,(6)) \ge 0.89$

4.6 Conclusion

In this paper, we have presented an approach for quantitative verification of complex stochastic systems such as driver assistant systems with a high dependence on the environmental situation. To this end, first the environmental situations together with the intended and unintended behaviour are characterized. In order to facilitate the elicitation of such requirements, we propose a visual specification in combination with additional formal requirements. As these specifications can be translated into a formal requirement, we can combine them with white box knowledge about the system under test for a subsequent verification step. As a formal basis for such logical specification, we built upon existing logics of SSMT, which is sufficiently expressive to state safety properties of stochastic hybrid systems, thereby addressing a rich class of systems. Although we believe that this class of systems is capable of characterizing all possible behaviours to be observed in traffic situations, it still requires an abstract specification of these dynamics. While a partly visual specification language can certainly help to specify relevant scenarios, a more detailed investigation of the capabilities of such specification language is subject to future research.

Using statistical model checking, we could then generate quantitative statements about the performance of the system with respect to a criticality measure. However, in a more practical situation, such white box knowledge might not be available, hence we had to adapt our method to be applicable in these black box situations as well. As this relaxes the assumptions, we cannot expect the safety statements to be of the same quality. In fact, the statistical guarantee is not only of a different type but also when constructing uncertainty intervals, these are considerably larger, and hence require significant more simulation runs for high safety targets. However, in contrast to the white box method, the black box method scales linearly with the number of simulation runs. As a rough comparison, we observed for our simulations an overall runtime for the white box method of tens of hours, whereas the black box method provided results within an hour. Nevertheless, it provides a solid baseline for further improvements concerning efficiency as well as usability with respect to the visual elicitation of scenario-based requirements.

References

1. Bartels, A., Meinecke, M.M., Steinmeyer, S.: Fahrstreifenwechselassistenz. In: Handbuch Fahrerassistenzsysteme, pp. 959–974. Springer, Berlin (2015)
2. Bubeck, S., Munos, R., Stoltz, G., Szepesvari, C.: X-armed bandits. J. Mach. Learn. Res. **12**(May), 1655–1695 (2011)
3. Damm, W., Harel, D.: LSCs: breathing life into message sequence charts. Formal Methods Syst. Des. **19**(1), 45–80 (2001)
4. Damm, W., Kemper, S., Möhlmann, E., Peikenkamp, T., Rakow, A.: Traffic Sequence Charts - From Visualization to Semantics. Reports of SFB/TR 14 AVACS 117, SFB/TR 14 AVACS (2017). http://www.avacs.org/fileadmin/Publikationen/Open/avacs_technical_report_117.pdf, http://www.avacs.org
5. Damm, W., Kemper, S., Mhlmann, E., Peikenkamp, T., Rakow, A.: Traffic Sequence Charts - A Visual Language for Capturing Traffic Scenarios. In: Embedded Real Time Software and Systems - ERTS2018 (2018). To appear
6. Damm, W., Mhlmann, E., Peikenkamp, T., Rakow, A.: A formal semantics for traffic sequence charts. In: Festschrift in honor of Edmund A. Lee (2017). To appear
7. David, A., Jensen, P.G., Larsen, K.G., Mikučionis, M., Taankvist, J.H.: Uppaal stratego. In: International Conference on Tools and Algorithms for the Construction and Analysis of Systems, pp. 206–211. Springer, Berlin (2015)
8. Ellen, C., Gerwinn, S., Fränzle, M.: Statistical model checking for stochastic hybrid systems involving nondeterminism over continuous domains. Int. J. Softw. Tools Technol. Transfer **17**(4), 485–504 (2015)
9. ENABLE-S[3]: (European initiative to enable validation for highly automated safe and secure systems). Grant nr. 692455-2 Call H2020-ECSEL-2015-2-IA-two-stage call (2016)
10. Fränzle, M., Hermanns, H., Teige, T.: Stochastic satisfiability modulo theory: a novel technique for the analysis of probabilistic hybrid systems. In: International Workshop on Hybrid Systems: Computation and Control, pp. 172–186. Springer, Berlin (2008)
11. Henzinger, T.A.: The theory of hybrid automata. In: Proceedings, 11th Annual IEEE Symposium on Logic in Computer Science, New Brunswick, NJ, USA, July 27–30, 1996, pp. 278–292. IEEE Computer Society (1996). https://doi.org/10.1109/LICS.1996.561342
12. Kemper, S., Etzien, C.: A visual logic for the description of highway traffic scenarios. In: Complex Systems Design & Management, pp. 233–245. Springer, Berlin (2014)
13. Linker, S., Hilscher, M.: Proof theory of a multi-lane spatial logic. In: International Colloquium on Theoretical Aspects of Computing, pp. 231–248. Springer, Berlin (2013)
14. Loos, S.M., Platzer, A., Nistor, L.: Adaptive cruise control: hybrid, distributed, and now formally verified. In: International Symposium on Formal Methods, pp. 42–56. Springer, Berlin (2011)
15. Nidhi Kalra, S.M.P.: Driving to Safety: How Many Miles of Driving Would It Take to Demonstrate Autonomous Vehicle Reliability? RAND Corporation (2016). http://www.jstor.org/stable/10.7249/j.ctt1btc0xw

16. PEGASUS: (projekt zur etablierung von generell akzeptierten gütekriterien, werkzeugen und methoden sowie szenarien und situationen zur freigabe hochautomatisierter fahrfunktionen). Funding: BMWI, Germany (2016)
17. Shmarov, F., Zuliani, P.: Probreach: verified probabilistic delta-reachability for stochastic hybrid systems. In: Proceedings of the 18th International Conference on Hybrid Systems: Computation and Control, pp. 134–139. ACM, New York (2015)
18. Stellet, J.E., Zofka, M.R., Schumacher, J., Schamm, T., Niewels, F., Zöllner, J.M.: Testing of advanced driver assistance towards automated driving: a survey and taxonomy on existing approaches and open questions. In: 2015 IEEE 18th International Conference on Intelligent Transportation Systems (ITSC), pp. 1455–1462. IEEE, New York (2015)
19. Vidyasagar, M.: Randomized algorithms for robust controller synthesis using statistical learning theory. Automatica **37**(10), 1515–1528 (2001)
20. Winner, H., Weitzel, A.: Quo vadis, fas? In: Handbuch Fahrerassistenzsysteme, pp. 658–667. Springer, Berlin (2012)
21. Younes, H.L., Kwiatkowska, M., Norman, G., Parker, D.: Numerical versus statistical probabilistic model checking. Int. J. Softw. Tools Technol. Transfer **8**(3), 216–228 (2006)

Chapter 5
Game Theory-Based Traffic Modeling for Calibration of Automated Driving Algorithms

Nan Li, Mengxuan Zhang, Yildiray Yildiz, Ilya Kolmanovsky and Anouck Girard

Abstract Automated driving functions need to be validated and calibrated so that a self-driving car can operate safely and efficiently in a traffic environment where interactions between it and other traffic participants constantly occur. In this paper, we describe a traffic simulator capable of representing vehicle interactions in traffic developed based on a game-theoretic traffic model. We demonstrate its functionality for parameter optimization in automated driving algorithms by designing a rule-based highway driving algorithm and calibrating the parameters using the traffic simulator.

5.1 Introduction

Significant efforts have been put into research, development, and implementation of automated driving algorithms, aiming at making self-driving cars a viable option for everyday transportation, for example, see [2, 16]. Several challenges and issues in

---

N. Li (✉) · M. Zhang · I. Kolmanovsky · A. Girard
Department of Aerospace Engineering, University of Michigan,
1320 Beal Avenue, Ann Arbor, MI 48109, USA
e-mail: nanli@umich.edu

M. Zhang
e-mail: mengxuan@umich.edu

I. Kolmanovsky
e-mail: ilya@umich.edu

A. Girard
e-mail: anouck@umich.edu

Y. Yildiz
Department of Mechanical Engineering, Bilkent University, 06800 Ankara, Turkey
e-mail: yyildiz@bilkent.edu.tr

© Springer International Publishing AG, part of Springer Nature 2019
H. Waschl et al. (eds.), *Control Strategies for Advanced Driver Assistance Systems and Autonomous Driving Functions*, Lecture Notes in Control and Information Sciences 476, https://doi.org/10.1007/978-3-319-91569-2_5

technical, legal, and social aspects are faced and must be addressed by automated driving researchers to achieve this goal—some of these challenges and issues are summarized in [8, 11].

To validate an automated driving algorithm and calibrate its parameters, hundreds of thousands of miles of driving tests may be required to cover a sufficiently diverse set of traffic scenarios and road conditions [1]. Consequently, simulations, model-based developments, and hardware-in-the-loop techniques need to be relied upon to reduce the need for extensive road testing and to maintain short time-to-market for automated driving technologies.

Traffic simulators can facilitate the initial calibration of automated driving algorithms before actual road tests and reduce the time and effort needed for the overall algorithm testing. As an example, in [18], a virtual environment for testing advanced driver assistance systems is proposed, which is configured using real-world driving data.

Since vehicle operation in traffic is inherently interactive, that is, the actions of car drivers are influenced by and also influence the actions of other traffic participants, a simulator used for testing and discovering faults in automated driving algorithms must reflect such interactions. In [9, 10], we have developed a simulator capable of representing vehicle interactions in traffic based on a hierarchical reasoning theory and level-k games. We have illustrated the use of the simulator to support quantitative analyses and comparisons of various automated driving decision and control systems, and support their initial calibrations. In this paper, we focus on the demonstration of the simulator's functionality for parameter optimization in automated driving algorithms. As a specific case study, a rule-based algorithm for automated highway driving is considered and its parameters are calibrated using the developed traffic simulator.

In the level-k game theory, each player in the game is assumed to have bounded rationality, and his/her level of rationality is represented by a finite reasoning depth k, see [15]. This assumption is arguably representative of human interactions in real-world situations and is supported by experimental evidence, see also [3–5]. The level-k game theory-based framework is particularly useful to model the interactive behavior of multiple players in a game when the game is complex, which is the case in multi-vehicle traffic scenarios. Similar in spirit implementations have been reported for modeling human pilot-to-unmanned aerial vehicle interactions in [13, 17]. In this paper, the traffic on a highway with n lanes, $n \in \mathbb{N}_+$, composed of up to 30 interactive vehicles, is modeled.

This paper is organized as follows: In Sect. 5.2, we describe our highway traffic modeling approach exploiting the level-k game theory. In Sect. 5.3, we introduce the simulator developed based on the proposed game-theoretic traffic model. In Sect. 5.4, we describe a rule-based automated highway driving algorithm to be tested and calibrated. In Sect. 5.5, we present an illustration of the use of the simulator to calibrate algorithm parameters. Finally, Sect. 5.6 presents a summary and concluding remarks.

5.2 Traffic Modeling Based on Level-k Game Theory

The traffic we are modeling in this work is n-lane highway traffic. Each car in the traffic can cruise with a steady speed, accelerate to a higher speed, decelerate to a lower speed, and change lanes to overtake other cars. Each car is assumed to be controlled by a human driver who obeys the general traffic rules and pursues safe, efficient, and comfortable travel.

5.2.1 Modeling of a Single Car

A car (including the driver) is modeled as a hierarchical system composed of two levels of controllers. A higher-level controller, that represents the human driver's decision-making processes, selects a maneuver (that in this paper is called an "action," denoted by γ) from a finite action set, according to the current state of the car in the traffic, to execute at each time step. Then a lower-level controller controls the engine, powertrain and vehicle dynamics according to the action command from the higher-level controller to let the car realize the prescribed maneuver. The structure of the system is shown in Fig. 5.1. In this work, we focus on the modeling of the higher-level decision making. The lower-level control and the car dynamics are represented by the following discrete-time dynamics,

$$\begin{aligned} x(t+1) &= x(t) + v_x(t)\, \Delta t, \\ v_x(t+1) &= v_x(t) + a(t)\, \Delta t, \\ y(t+1) &= y(t) + v_y(t)\, \Delta t, \end{aligned} \quad (5.1)$$

where x and v_x represent, respectively, the position and velocity of the car on the highway in the longitudinal direction, while y represents the position of the car on the highway in the lateral direction. The longitudinal acceleration, $a(t)$, and the lateral velocity, $v_y(t)$, are two control inputs that are decided by the higher-level controller corresponding to the selected action commands.

In traffic, a driver can only observe and process a limited amount of information, due to the limits of human vision and the human brain. In particular, a driver can observe the motion of the cars in an immediate vicinity of his/her own and decide on his/her next maneuver based on these observations. We assume that the following observations are used by a driver in his/her decision making, as supported by [7, 12]:

- The range and range rate to the car in front and in the same lane, d_{fc} and v_{fc},
- The range and range rate to the car in front and in the left lane, d_{fl} and v_{fl},
- The range and range rate to the car in front and in the right lane, d_{fr} and v_{fr},
- The range and range rate to the car in the rear and in the left lane, d_{rl} and v_{rl},
- The range and range rate to the car in the rear and in the right lane, d_{rr} and v_{rr},
- The lane index of the ego car, $i \in \{1, 2, \ldots, n\}$,

Fig. 5.1 Control hierarchy of the car model

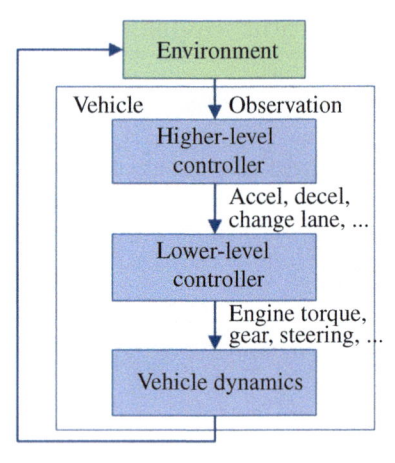

where "range" is defined as the longitudinal distance from the ego car to another car.

Therefore, the observation state of a car is defined by a vector

$$x_{ob} = [d_{fl}\ v_{fl}\ d_{fc}\ v_{fc}\ d_{fr}\ v_{fr}\ d_{rl}\ v_{rl}\ d_{rr}\ v_{rr}\ i]^\top \in \mathbb{R}^{11}. \qquad (5.2)$$

It is noted that a human driver may not be able to accurately measure his/her range or range rate to another car. He/she can only estimate and specify them, respectively, as "close," "far," and as "approaching," "moving away," etc. In this paper, we introduce categorical values to encode the ranges as follows:

$$Spec(d) = \begin{cases} \text{"close"} & \text{if } d \leq d_c, \\ \text{"nominal"} & \text{if } d_c < d \leq d_f, \\ \text{"far"} & \text{if } d > d_f, \end{cases} \qquad (5.3)$$

and introduce categorical values to encode the range rates as follows:

$$Spec(v) = \begin{cases} \text{"approaching"} & \text{if } v < 0, \\ \text{"stable"} & \text{if } v = 0, \\ \text{"moving away"} & \text{if } v > 0. \end{cases} \qquad (5.4)$$

We remark that the observation states and encoding levels must be introduced judiciously. More states and levels may improve the fidelity of the car model but can impede the computations and the implementation due to the "curse of dimensionality."

The action set, denoted by Γ, covers the major maneuver actions a human driver normally applies in highway traffic:

- "Maintain" current lane and speed: $a = 0, v_y = 0$;
- "Accelerate" at a nominal rate: $a = a_1, v_y = 0$;
- "Decelerate" at a nominal rate: $a = -a_1, v_y = 0$;

- "Hard accelerate" at a large rate: $a = a_2, v_y = 0$;
- "Hard decelerate" at a large rate: $a = -a_2, v_y = 0$;
- Change lane "to the Left": $a = 0, v_y = v_{cl}$;
- Change lane "to the Right": $a = 0, v_y = -v_{cl}$.

The nominal rate, $\pm a_1$ (m/s^2), reflects the acceleration/deceleration a human driver would apply in normal situations, while the "hard" rate, $\pm a_2$ (m/s^2), reflects the acceleration/deceleration a human driver would apply in aggressive driving situations. The latter values depend on the maximum acceleration/deceleration capability of a car.

Additional modeling assumptions are made as follows: The longitudinal velocity of a car is assumed to be bounded to a range $v_x \in [v_{\min}, v_{\max}]$, and is saturated to this range during the simulations. The lateral velocity of a car during a lane change is assumed to be constant such that the total time to change lanes is t_{cl}, i.e.,

$$v_{cl} = \frac{w}{t_{cl}}, \quad (5.5)$$

where w is the width of a lane. Also, we assume that a car always drives at the center of a lane unless the car is performing a lane change. Once a lane change begins, it always continues to completion.

The model of a driver is a map from the observations to the actions, also called a "policy." The driver selects an action, $\gamma^* \in \Gamma$, to execute based on his/her current observation state, to pursue his/her driving goals at each time step. Basic goals of a driver in highway traffic are (1) to maintain sufficient separation with other vehicles and avoid involvement in accidents, such as car crashes (safety), (2) to minimize the time needed to reach his/her destination (performance), (3) to keep a reasonable headway from preceding cars (safety and comfort), and (4) to minimize driving effort (comfort).

The pursuit of these goals can be reflected in a reward function that is to be maximized. The reward function, in this paper, is designed as

$$R = w_1 \hat{c} + w_2 \hat{v} + w_3 \hat{h} + w_4 \hat{e}, \quad (5.6)$$

where w_i, $i = 1, 2, 3, 4$, are the weights for each term and $\hat{c}, \hat{v}, \hat{h}$, and $\hat{e}$ represent "(safety) constraint violation," "(longitudinal) velocity," "headway," and "effort" metrics, respectively. The weights, w_i, may change depending on the aggressiveness of the driver; however, the safety typically has the highest priority. Hence, the following relationship between the weights should be kept:

$$w_1 \gg w_2, w_3, w_4. \quad (5.7)$$

The terms, $\hat{c}, \hat{v}, \hat{h}, \hat{e}$, are explained further below.

$\hat{c}$ (**constraint violation**): We define a safe zone for each car (a rectangular area that over-bounds the geometric contour of the car with a safety margin) whose boundaries are treated as safety constraints. The term $\hat{c}$ is assigned a value of -1 when a constraint violation occurs, that is, the safe zone is invaded by another car, and a value of 0 otherwise.

$\hat{v}$ (**velocity**): The term $\hat{v}$ is assigned the value

$$\hat{v} = \frac{v_x - v_{\text{nominal}}}{a_1}, \quad v_{\text{nominal}} = \frac{v_{\min} + v_{\max}}{2}. \tag{5.8}$$

Dividing by a_1 makes this term of the same order of magnitude as other terms and facilitates the design of the weights.

$\hat{h}$ (**headway**): The term $\hat{h}$ takes the following values depending on the headway distance (the range to the car directly in front):

$$\hat{h} = \begin{cases} -1 & \text{if } Spec(d_{\text{fc}}) = \text{"close,"} \\ 0 & \text{if } Spec(d_{\text{fc}}) = \text{"nominal,"} \\ 1 & \text{if } Spec(d_{\text{fc}}) = \text{"far."} \end{cases} \tag{5.9}$$

$\hat{e}$ (**effort**): The term $\hat{e}$ takes the value 0 if the driver's action is "Maintain," $\hat{e} = \hat{e}_h$ if the driver's action is "Hard accelerate" or "Hard decelerate," and $\hat{e} = \hat{e}_n$ otherwise. This term discourages the driver from making unnecessary maneuvers. In particular, if the driver is able to maintain safety with other actions, a higher penalty would discourage the driver from applying "Hard accelerate" or "Hard decelerate," that decrease the comfort. But in the case where another maneuver cannot avoid constraint violation, the driver would apply "Hard accelerate" or "Hard decelerate" to maintain safety. Note that the ratio between $\hat{e}_n$ and $\hat{e}_h$ has an influence on the driver's behavior and may be adjusted to match the driving behavior of different human drivers. In this paper, we choose $\hat{e}_n = -1$ and $\hat{e}_h = -5$.

In the above reward function (5.6), we impose a penalty on constraint violations. We also note that there are some combinations of observation states and actions that obviously lead to undesired behaviors, such as constraint violations. We can impose ancillary constraints to avoid the occurrence of such combinations. The imposition of such constraints also benefits the numerical convergence of the reinforcement learning algorithm, that is, applied to compute the optimal control policy. These ancillary constraints are as follows:

- If a car in the left lane is in a parallel position, the ego car cannot select the "to the Left" action,
- If a car in the right lane is in a parallel position, the ego car cannot select the "to the Right" action,
- If $Spec(d_{\text{fl}}) = $ "close" and $Spec(v_{\text{fl}}) = $ "approaching," or, $Spec(d_{\text{rl}}) = $ "close" and $Spec(v_{\text{rl}}) = $ "approaching," the ego car cannot select the "to the Left" action,
- If $Spec(d_{\text{fr}}) = $ "close" and $Spec(v_{\text{fr}}) = $ "approaching," or, $Spec(d_{\text{rr}}) = $ "close" and $Spec(v_{\text{rr}}) = $ "approaching," the ego car cannot select the "to the Right" action.

Two cars are assumed to be in a "parallel" position when their safe zones intersect in the longitudinal direction.

We remark that when selecting actions, the ego driver is supposed to not only consider his/her current observation state, but also consider the potential actions of the other cars in his/her vicinity, as their actions influence the evolution of the ego car's states. This reflects the interactive nature of traffic dynamics. In this work, we exploit a hierarchical reasoning and level-k game theory-based approach to model the vehicle-to-vehicle interactions in highway traffic.

5.2.2 Modeling of Interactive Traffic

In this work, the scheme for modeling interactive highway traffic is premised on the assumptions that (1) when a strategic agent makes its own decisions in a multi-agent interactive scenario (such as a traffic scenario composed of multiple cars), it takes into consideration the predicted decisions of the other agents; (2) an agent predicts the decisions of the other agents through a finite depth of reasoning (called "level"), and different agents may have different reasoning levels. The reader is referred to [4, 15] for more comprehensive discussions on hierarchical reasoning and level-k game theory.

To model the drivers of different reasoning levels, one starts by specifying a "nonstrategic" driver model, which is referred to as a "level-0" driver model. A "level-0" driver makes instinctive action decisions to pursue his/her own goals without considering the potential actions or reactions of the other drivers. Then, we use a reinforcement learning algorithm to determine the model of a "level-1" driver; that is, the optimal observation-to-action map (policy) based on the reward function (5.6) and assuming that all of the drivers in the traffic but him/herself are level-0 drivers. Similarly, a level-k driver takes optimal actions assuming that all of the other drivers react as level-$(k-1)$ drivers. It is suggested by experimental studies in [4] that humans are usually level-0, 1, or 2 reasoners in their interactions. So we generate driver models up to level-2. In general, our traffic simulator can be configured based on certain fractions of drivers of each level. Different reasoning levels may reflect different driving habits and proficiency levels of different human drivers. In principle, the reward functions of the drivers may also not need to be the same.

In this scheme, the underlying dynamics of the traffic composed of multiple cars are modeled as a Markov decision process, whose state is determined by the states (x, v_x, y) of all cars. In particular, as discussed in the previous subsection, a car only obtains a limited amount of information from its vicinity through its observations, i.e., the whole state of the traffic is only partially observable to each car. Hence, determining a control policy in this setting is a partially observable Markov decision process (POMDP) problem. In this work, as in [10], we employ the Jaakkola reinforcement learning (RL) algorithm [6], that distinguishes itself from conventional approaches by guaranteeing convergence to a local maximum in terms of average rewards when the problem is of POMDP type and states and actions admit a finite number of values, to solve for the level-k driver models.

Fig. 5.2 Reinforcement learning algorithm to obtain the level-k driver models

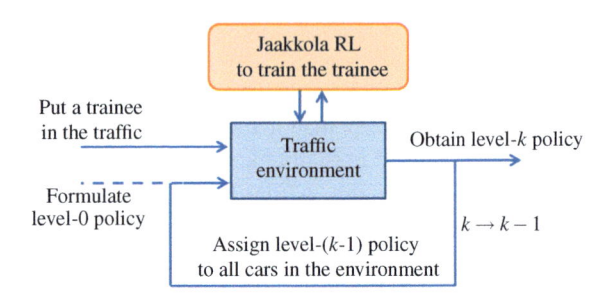

Figure 5.2 shows the procedures to obtain the level-k driver models using the Jaakkola RL algorithm. To obtain the level-k policy, one assigns the level-$(k-1)$ policy to all of the drivers in the traffic except for the driver being trained (trainee), lets the trainee interacts with those level $(k-1)$ drivers and exploits the Jaakkola RL algorithm to gradually improve the trainee's policy. The optimal policy, that assigns a probability distribution over all possible actions to each observation state, is expected to gain the highest average reward if it is executed over an infinitely long simulation. Through Jaakkola RL, the optimal policy is obtained when the average reward converges during the training. For more details on the Jaakkola RL algorithm and its implementation to obtain the level-k policies, see [10].

The development of level-k policies starts by specifying a "level-0" policy. In this paper, we formulate the level-0 policy by prescribing minimally rational behaviors for a range of observation states, as follows:

$$\gamma_{i0}^* = \begin{cases} \text{"Decelerate,"} & \text{if } Spec(d_{\text{fc}}) = \text{"nominal"} \ \& \ Spec(v_{\text{fc}}) = \text{"approaching,"} \\ & \text{or } Spec(d_{\text{fc}}) = \text{"close"} \ \& \ Spec(v_{\text{fc}}) = \text{"stable,"} \\ \text{"Hard decelerate,"} & \text{if } Spec(d_{\text{fc}}) = \text{"close"} \ \& \ Spec(v_{\text{fc}}) = \text{"approaching,"} \\ \text{"Maintain,"} & \text{otherwise.} \end{cases}$$

(5.10)

5.3 Traffic Simulator Based on Level-k Game Theory

The simulator has been configured with vehicles responding according to level-k driving policies to represent traffic on a 2-lane highway and a 3-lane highway. Note that we have considered case studies for a 3-lane highway also in previous publications [9, 10, 14], but have not reported the results for a 2-lane highway case.

Figure 5.3 shows snapshots of a 2-lane highway traffic simulation and of a 3-lane highway traffic simulation. In the figures, the red car in the center is the ego car—it can be controlled by an automated driving algorithm being tested; the yellow cars constitute the traffic environment and all of them are modeled using the level-k

Fig. 5.3 Traffic simulator. **a** Simulation of a traffic scenario on a 2-lane highway. **b** Simulation of a traffic scenario on a 3-lane highway

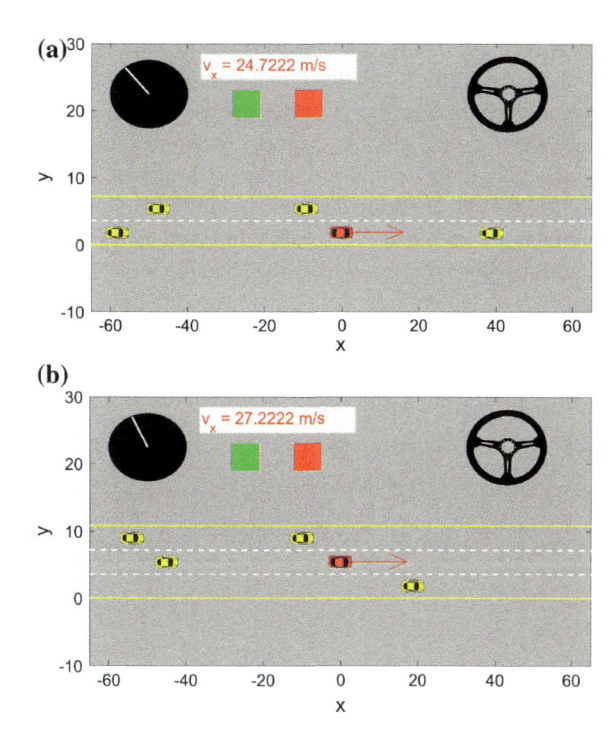

policies—they interact with each other as well as with the red automated car. The red arrow attached to the red car indicates its travel direction and arrow length indicates the car's travel speed. On the left is a speedometer, and on the right the steering wheel indicates the lateral motion of the red car. The green box and red box in the middle indicate, respectively, the gas pedal and the brake pedal.

We evaluate the level-k game-theoretic driver models by letting the red test car be controlled by a level-k policy and testing it in a level-$(k-1)$ traffic environment (all of the yellow cars are modeled using level-$(k-1)$ policies). When training the level-k policies, we use $w_1 = 10,000$, $w_2 = 5$, $w_3 = 1$, $w_4 = 1$. We remark that we use a larger weight w_2 for travel speed metric $\hat{v}$ compared to the weights w_3, w_4 for headway distance metric $\hat{h}$ and for driving effort metric $\hat{e}$; as a result, the obtained policy is of increased aggressiveness—tending to make more frequent maneuvers for overtaking other cars in order to achieve higher travel speeds. This setup is synergistic with the needs for testing automated driving control algorithms, because it provides more complex and challenging scenarios. We use constraint violation rate and average travel speed as metrics to quantitatively evaluate the safety and performance of the level-k policies. Here, "constraint violation" refers to the test car's safe zone being entered by any of the cars in the simulations. To obtain these rates, 10,000 simulations are run for each number of cars (that reflects the traffic density). Each simulation is 200[s] long and the rates are provided as the percentage of simulation runs during which at least one "constraint violation" occurs.

We remark that for the results presented in Figs. 5.4 and 5.5, the level-2 policy is evaluated in a traffic environment composed of level-1 cars, and the level-1 policy is evaluated in a traffic environment composed of level-0 cars. It is observed that the level-2 policy exhibits higher constraint violation rates than the level-1 policy. One explanation for this is that the dynamics of the level-1 traffic where the level-2 policy is tested are more aggressive and harder to predict because of the complicated interactions simultaneously happening among the level-1 cars and between the level-1 and the level-2 cars; while, the dynamics of the level-0 traffic where the level-1 policy is tested are relatively easy to predict due to the non-strategic behavior of the level-0 cars. Also, it is observed that the level-2 car has higher average travel speed compared to the level-1 car. One explanation for this is that the flow speed of the level-1 traffic is, on average, higher than that of the level-0 traffic, because the level-

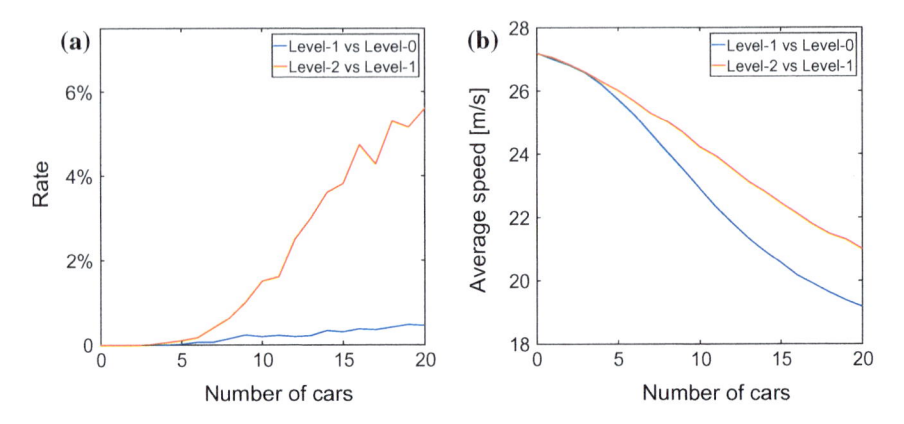

Fig. 5.4 Simulation results for level-k cars driving on a 2-lane highway: **a** Constraint violation rate. **b** Average travel speed

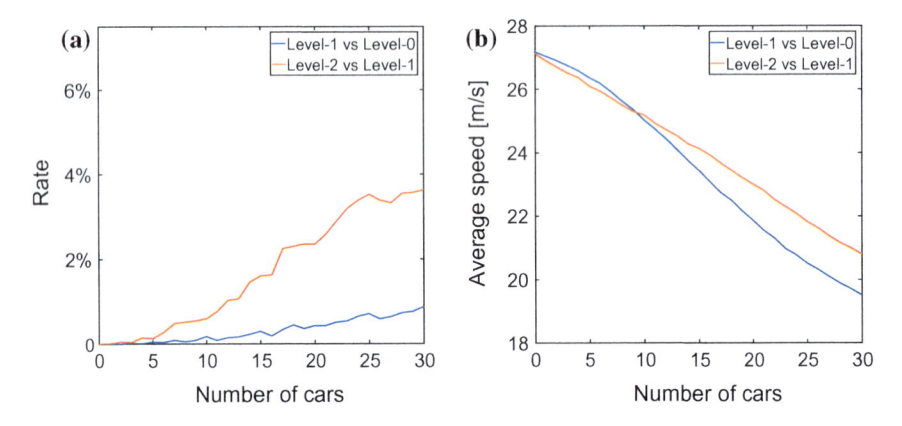

Fig. 5.5 Simulation results for level-k cars driving on a 3-lane highway: **a** Constraint violation rate. **b** Average travel speed

1 cars can change lanes to overtake slower cars while the level-0 cars cannot. As a consequence, the level-2 car in the level-1 traffic flow also has more possibilities to travel faster, compared to the level-1 car driving in the level-0 traffic flow, that may block the level-1 car from traveling faster. We note that these observations hold for both a 2-lane highway and a 3-lane highway traffic.

We then consider a traffic model composed of a mix of cars operating with different level-k policies. Specifically, in accordance with the results of an experimental study conducted in [4], we consider a traffic model where 10% of the drivers make decisions based on level-0 policies, 60% of the drivers act based on level-1 policies and 30% use level-2 policies. This traffic model is used hereafter to evaluate automated driving control algorithms and support their optimal parameter calibration.

We remark that the level-k cars in the traffic model utilize the 11 observations (5.2) to determine their actions from the discrete action set Γ, as described in Sect. 5.2.1; the test car can use a different observation set (referred to as the "sensing system") and a different control set that are decided by the algorithm developer to achieve a good balance between performance and complexity. In the rule-based automated highway driving algorithm developed in the next section, we utilize the same sensing system (5.2).

5.4 Rule-Based Automated Highway Driving Algorithm

To illustrate the use of the simulator for optimal parameter calibration of automated driving algorithms, we consider a specific rule-based highway driving algorithm, that includes both the longitudinal motion control and the lateral lane change decision making. We optimize its parameters based on an objective function that reflects both safety and performance requirements.

The proposed automated highway driving algorithm is a finite state machine (FSM) that has three states (also called "modes"), see Fig. 5.6. In each mode, the vehicle's longitudinal motion and lateral motion are controlled based on specific laws; mode switches are triggered when certain conditions get satisfied.

Now we introduce the control laws for each mode and the switch conditions between each pair of modes. In the sequel, the mode of the car at time t is denoted by $M(t)$; the p_i, $i = 1, \ldots, 6$, are boolean variables and each of them denotes the truth value of a specific condition.

5.4.1 Cruise Control Mode - C

When in the cruise control mode (C), the car maintains a reference speed v_{ref} by adjusting the acceleration according to

$$a(t) = K_c \big(v_{\text{ref}} - v_x(t)\big), \tag{5.11}$$

Fig. 5.6 Finite state machine diagram for automated highway driving. The labels [1–9] designate switch conditions between different modes

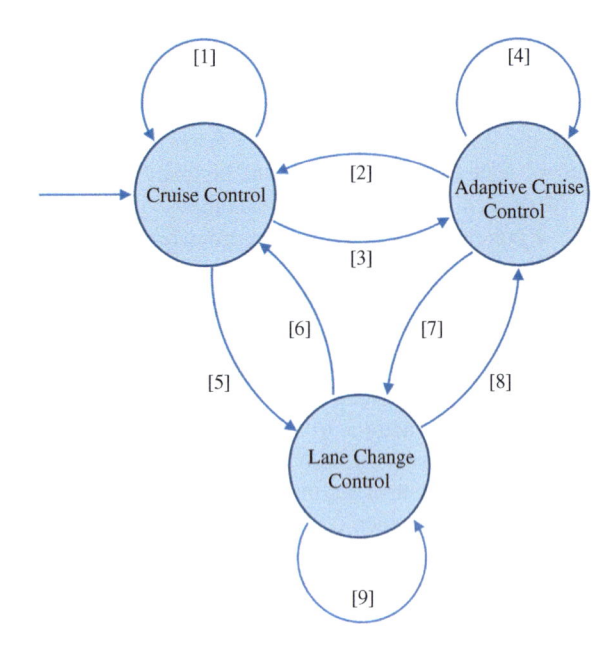

where K_c is the gain to match the car's actual speed $v_x(t)$ to a reference speed v_{ref}. In the cruise control mode $M(t) = C$, allowable mode switches are:

$$M(t+1) = \begin{cases} A & \text{if } p_1, \\ C & \text{if } \neg p_1, \end{cases} \quad (5.12)$$

where p_1 denotes the truth value of the condition

$$p_1 : d_{\text{fc}} \leq d_{\text{acc}},$$

where d_{acc} represents a critical headway distance. Under the above condition, the car switches from the cruise control mode to the adaptive cruise control mode (A) to perform car following when there is a car directly in front within this critical distance.

5.4.2 Adaptive Cruise Control Mode - A

When in the adaptive cruise control mode (A), the car follows a leader (l) while keeping a desired distance d_{des} to the leader by adjusting the acceleration according to

$$a(t) = K_p\left(x^l(t) - x(t) - d_{\text{des}}\right) + K_v\left(v_x^l(t) - v_x(t)\right), \quad (5.13)$$

where K_p is the gain to match the actual car-following distance $x^l(t) - x(t)$ to a desired distance d_{des}, and K_v is the gain to match the ego car's speed $v_x(t)$ to the speed of the leader $v_x^l(t)$. The acceleration policy (5.13) can alternatively be defined in terms of time headway.

The acceleration/deceleration of the car is assumed to be bounded by $a(t) \in [-a_2, a_2]$ (if the computed $a(t)$ is outside the bounds, it is saturated to the bounds). In the adaptive cruise control mode $M(t) = A$, allowable mode switches are:

$$M(t+1) = \begin{cases} C & \text{if } p_2, \\ L & \text{if } (\neg p_2) p_3 p_4 p_5, \\ A & \text{otherwise,} \end{cases} \quad (5.14)$$

where

p_2: $d_{\text{fc}} \geq d_{\text{cc}}$,
p_3: the predicted acceleration that can be obtained in the target lane
is larger than the acceleration obtained in the current lane,
p_4: $d_{\text{fc}} \geq d_{\text{win}}$,
p_5: ($d_{\text{fl}} \geq d_{\text{win}}$ and $d_{\text{rl}} \geq d_{\text{win}}$) or ($d_{\text{fr}} \geq d_{\text{win}}$ and $d_{\text{rr}} \geq d_{\text{win}}$),

where d_{cc} represents a critical headway distance such that when the car directly in front moves away and gets outside this critical distance, the car switches to cruise control mode to maintain the reference speed v_{ref}, and d_{win} represents a window of safe distance to complete a lane change.

The motivation for lane changes is for overtaking slower cars and pursuing higher travel speeds. When a predicted acceleration that can be achieved if the ego car travels in another lane is larger than the acceleration that is obtained by the ego car traveling in the current lane (p_3), the ego car may switch to the lane change mode (L) to perform a lane change. If only one of the left/right lanes satisfies the lane change condition ($\neg p_2) p_3 p_4 p_5$, that lane is set to be the target lane for lane change. If both satisfy the condition, the lane that leads to a larger acceleration is set to be the target lane.

5.4.3 Lane Change Mode - L

When in the lane change mode (L), the car changes lanes from the original lane to a target lane. In the lane change mode, the car's longitudinal speed remains constant and its lateral motion is determined by

$$v_y(t) = \pm \frac{w}{t_{\text{cl}}}, \quad (5.15)$$

where w is the width of a lane, t_{cl} is the time needed to make a lane change, and the plus/minus sign indicates the lane change direction. We assume that once a lane change begins, it always continues to completion.

In the lane change mode $M(t) = L$, allowable mode switches are:

$$M(t+1) = \begin{cases} A & \text{if } p_1 p_6, \\ C & \text{if } (\neg p_1) p_6, \\ L & \text{if } \neg p_6, \end{cases} \quad (5.16)$$

where

p_6: the car reaches the center of the target lane within a deviation tolerance.

5.4.4 Basic Simulation of Rule-Based Automated Highway Driving Controller

The operation of this rule-based automated highway driving controller in a simulation that is 50 [s] long is presented in Fig. 5.7. The traffic environment is modeled using the developed interactive traffic model composed of a mix of level-k cars, that has been discussed at the end of Sect. 5.3, on a 3-lane highway. Figure 5.7a shows the time history of the mode $M(t)$. Figure 5.7b shows the longitudinal speed of the car during the simulation. The parameter values used in the simulation are given below:

$$K_c = 0.25\,[\text{s}^{-1}], \quad K_p = 0.25\,[\text{s}^{-2}], \quad K_v = 1\,[\text{s}^{-1}],$$
$$v_{\text{ref}} = v_{\text{max}} = 98\,[\text{km/h}], \quad v_{\text{min}} = 62\,[\text{km/h}], \quad t_{cl} = 2\,[\text{s}], \quad (5.17)$$
$$d_{\text{acc}} = 37\,[\text{m}], \quad d_{\text{cc}} = 47\,[\text{m}], \quad d_{\text{des}} = 31.5\,[\text{m}], \quad d_{\text{win}} = 21\,[\text{m}].$$

Figure 5.8 quantitatively evaluates the performance of this rule-based controller in the modeled 3-lane highway traffic. The metrics are the same as those used to evaluate the level-k policies in Figs. 5.4 and 5.5. It is observed that the test car exhibits a significant number of constraint violations. On one hand, this may result from some deficiencies in the designed control strategy. For example, lack of some necessary observations, such as front cars' turning signals, may limit the test car's capability of handling uncertainties emanating from uncertain behaviors of other drivers. We remark that the level-k policies may not have such a problem because predictions of how neighbor cars behave have been taken into account implicitly during the RL training. On the other hand, the parameters of our rule-based controller may not be calibrated to the best values. The values of these parameters can be optimized with the support of the simulator as discussed next.

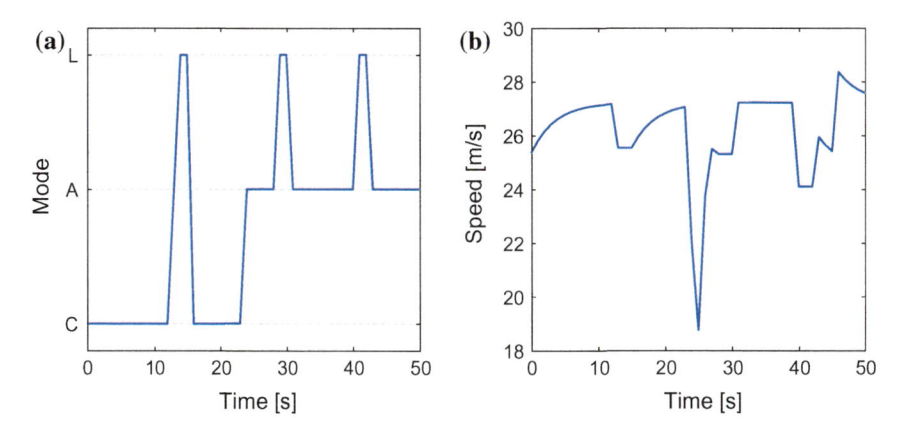

Fig. 5.7 The operation of the rule-based automated highway driving controller: **a** Mode history. **b** Travel speed history

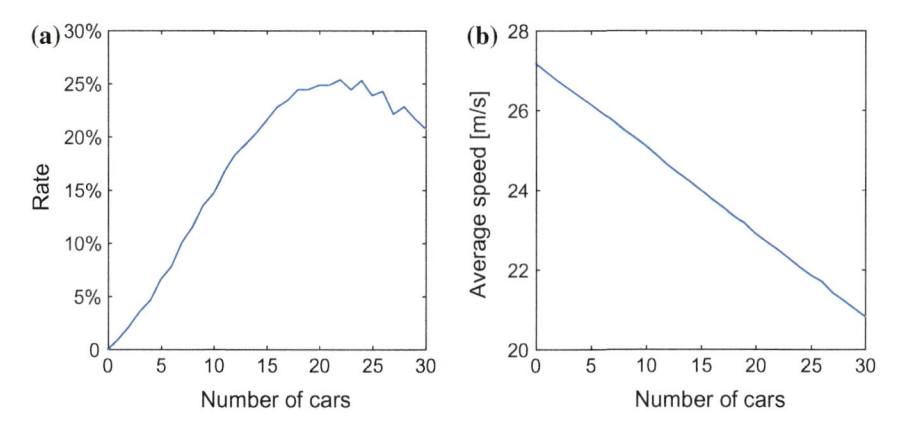

Fig. 5.8 Simulation results for the rule-based automated driving controller on a 3-lane highway: **a** Constraint violation rate. **b** Average travel speed

5.5 Optimal Automated Driving Controller Calibration

In this section, we calibrate the parameters of the rule-based automated highway driving controller developed in the previous section to improve performance, with the support of the simulator constructed based on the level-k game-theoretic traffic model. We remark that the approach discussed in this section can be used to calibrate other automated driving algorithms as well. Another example can be found in [10].

The performance of the controller can be represented by the value of a predefined objective function that accounts for safety, performance, comfort, fuel economy, etc. As an example, we define the objective function to be maximized as follows,

$$R_{obj} = k_1(-\bar{c}) + k_2 \frac{\bar{v}_x - v_{min}}{v_{max} - v_{min}}, \quad (5.18)$$

where the weights k_1 and k_2 are determined by the user, $\bar{c}$ is the constraint violation rate defined as in Figs. 5.4 and 5.5, $\bar{v}_x$ is the average travel speed during the simulations, and v_{min} and v_{max} represent, respectively, a speed lower bound and a speed upper bound for highway driving. We remark that (5.18) is designed such that each of its terms is dimensionless and normalized.

The parameters to be optimized in this example are: (1) the desired car-following distance d_{des} in (5.13), and (2) the safe distance window to make a lane change d_{win} in (5.14). The traffic scenario is modeled as 20 cars driving on a 3-lane highway. The value of (5.18) as a function of (d_{des}, d_{win}) is shown in Fig. 5.9. The numbers in the plots are obtained as the average objective function values over 1000 simulation runs for each pair of (d_{des}, d_{win}) values on a grid.

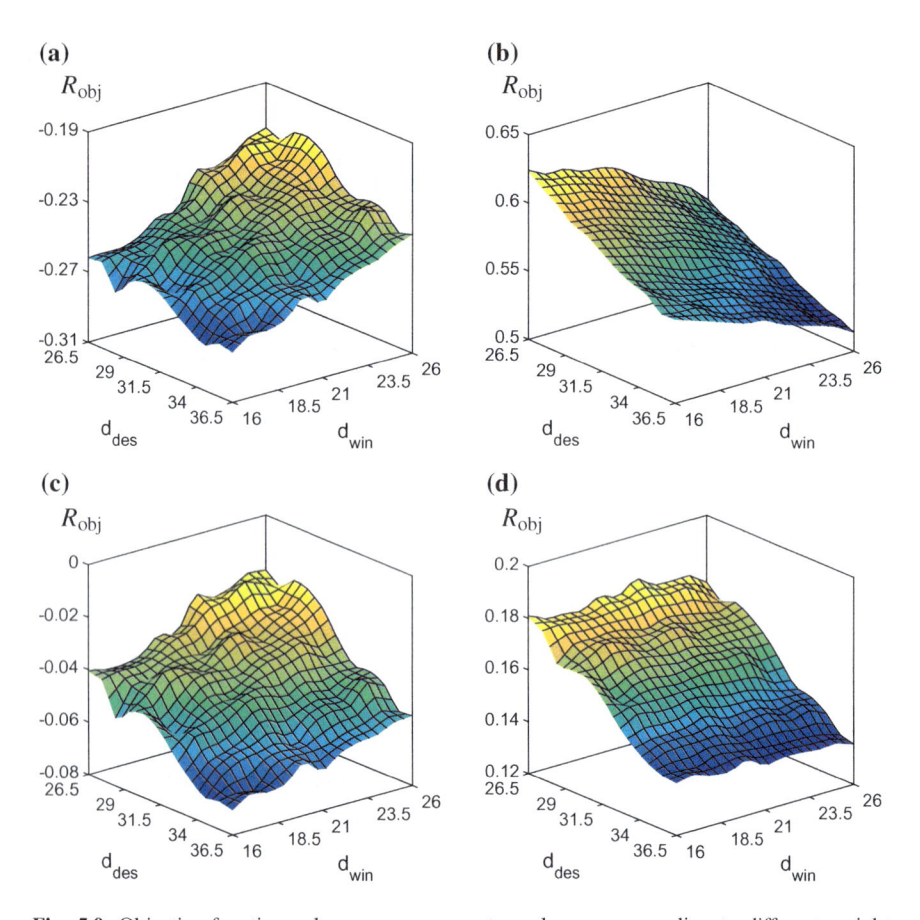

Fig. 5.9 Objective function values versus parameter values corresponding to different weight choices: **a** $k_1 = 1, k_2 = 0$. **b** $k_1 = 0, k_2 = 1$. **c** $k_1 = 0.75, k_2 = 0.25$. **d** $k_1 = 0.5, k_2 = 0.5$

Figure 5.9 can be used to select the best pair of (d_{des}, d_{win}) for a specific objective function design. For example, for maximum safety $k_1 = 1$ and $k_2 = 0$, it can be observed that the larger d_{win} is, the higher value one obtains, i.e., the safety is improved with a larger separation between vehicles. When both safety and speed are considered, for example, $k_1 = 0.75$, $k_2 = 0.25$, the best pair is (d_{des}, d_{win}) = (26.5, 25) (m). The other parameters of the algorithm, for example, K_c, K_p, K_v, v_{ref}, d_{des}, d_{acc}, d_{cc}, d_{win} and t_{cl}, can be optimized similarly.

5.6 Summary and Concluding Remarks

Simulators that account for interactions of vehicles in traffic can facilitate testing, validation, verification, and calibration of automated driving algorithms in the virtual world. They can also be used to uncover situations and scenarios that are particularly challenging for automated driving or are likely to result in faults in particular automated driving policies, thereby informing their future testing in the simulated or on-the-road conditions. Consequently, the extent of the required time and effort consuming on the road testing may potentially be reduced, thereby addressing one of the major current challenges in the automated vehicle development.

In this paper, we have described an approach to modeling vehicle interactions in traffic based on an application of the level-k game theory. Case studies of configuring the simulator to represent traffic on 2-lane and 3-lane highways and to evaluate and improve parameters in a rule-based automated highway driving policy have been presented. Some qualitative trends observed in the modeled traffic for different number of lanes and for different levels of vehicles involved have been presented. Future work will focus on extending our approach to represent city traffic.

Acknowledgements Nan Li and Ilya Kolmanovsky acknowledge the support of this research by the National Science Foundation under Award CNS 1544844 to the University of Michigan. Yildiray Yildiz acknowledges the support of this research by the Scientific and Technological Research Council of Turkey under Grant 114E282 to Bilkent University.

References

1. Anderson, J.M., Nidhi, K., Stanley, K.D., Sorensen, P., Samaras, C., Oluwatola, O.A.: Autonomous Vehicle Technology: A Guide for Policymakers. Rand Corporation (2014)
2. Campbell, M., Egerstedt, M., How, J.P., Murray, R.M.: Autonomous driving in urban environments: approaches, lessons and challenges. Philos. Trans. R. Soc. Lond. A Math. Phys. Eng. Sci. **368**(1928), 4649–4672 (2010)
3. Costa-Gomes, M.A., Crawford, V.P.: Cognition and behavior in two-person guessing games: an experimental study. Am. Econ. Rev. **96**(5), 1737–1768 (2006)
4. Costa-Gomes, M.A., Crawford, V.P., Iriberri, N.: Comparing models of strategic thinking in van huyck, battalio, and beil's coordination games. J. Eur. Econ. Assoc. **7**(2–3), 365–376 (2009)

5. Hedden, T., Zhang, J.: What do you think I think you think? Strategic reasoning in matrix games. Cognition **85**(1), 1–36 (2002)
6. Jaakkola, T., Singh, S.P., Jordan, M.I.: Reinforcement learning algorithm for partially observable markov decision problems. In: Advances in Neural Information Processing Systems, pp. 345–352. Citeseer (1995)
7. Kikuchi, S., Chakroborty, P.: Car-following model based on fuzzy inference system. Transp. Res. Rec. 82–82 (1992)
8. Langari, R.: Autonomous vehicles. In: 2017 American Control Conference (ACC), pp. 4018–4022 (2017). https://doi.org/10.23919/ACC.2017.7963571
9. Li, N., Oyler, D., Zhang, M., Yildiz, Y., Girard, A., Kolmanovsky, I.: Hierarchical reasoning game theory based approach for evaluation and testing of autonomous vehicle control systems. In: IEEE 55th Conference on Decision and Control (CDC), pp. 727–733. IEEE (2016)
10. Li, N., Oyler, D.W., Zhang, M., Yildiz, Y., Kolmanovsky, I., Girard, A.R.: Game theoretic modeling of driver and vehicle interactions for verification and validation of autonomous vehicle control systems. IEEE Trans. Control Syst. Technol. **99**, 1–16 (2017). https://doi.org/10.1109/TCST.2017.2723574
11. Maurer, M., Gerdes, J.C., Lenz, B., Winner, H.: Autonomous driving: Technical, Legal and Social Aspects. Springer Publishing Company, Incorporated (2016)
12. McDonald, M., Wu, J., Brackstone, M.: Development of a fuzzy logic based microscopic motorway simulation model. In: IEEE Conference on Intelligent Transportation System, pp. 82–87. IEEE (1997)
13. Musavi, N., Onural, D., Gunes, K., Yildiz, Y.: Unmanned aircraft systems airspace integration: a game theoretical framework for concept evaluations. J. Guidance Control Dyn. **40**(1), 96–109 (2016)
14. Oyler, D.W., Yildiz, Y., Girard, A.R., Li, N.I., Kolmanovsky, I.V.: A game theoretical model of traffic with multiple interacting drivers for use in autonomous vehicle development. In: 2016 American Control Conference (ACC), pp. 1705–1710 (2016)
15. Stahl, D.O., Wilson, P.W.: On players models of other players: theory and experimental evidence. Games Econ. Behav. **10**(1), 218–254 (1995)
16. Urmson, C., Anhalt, J., Bagnell, D., Baker, C., Bittner, R., Clark, M., Dolan, J., Duggins, D., Galatali, T., Geyer, C., et al.: Autonomous driving in urban environments: Boss and the urban challenge. J. Field Rob. **25**(8), 425–466 (2008)
17. Yildiz, Y., Agogino, A., Brat, G.: Predicting pilot behavior in medium-scale scenarios using game theory and reinforcement learning. J. Guidance Control Dyn. **37**(4), 1335–1343 (2014)
18. Zhou, J., Schmied, R., Sandalek, A., Kokal, H., del Re, L.: A framework for virtual testing of ADAS. SAE Int. J. Passeng. Cars Electron. Electr. Syst. **9**(2016-01-0049), 66–73 (2016)

Chapter 6
A Virtual Development and Evaluation Framework for ADAS—Case Study of a P-ACC in a Connected Environment

Harald Waschl, Roman Schmied, Daniel Reischl and Michael Stolz

Abstract Advanced driver assistance systems (ADAS) or even (partially) automated driving functions (ADF) can lead to substantial improvements in fuel economy, safety, and comfort of passenger cars. Especially, in view of new technologies, such as connected vehicles, additional improvements are feasible. However, testing and validation of ADAS in a connected and interacting environment are a critical and not yet fully solved task. In real-world driving situations in a dense urban traffic environment, constant interactions between the system under test (SUT) and other traffic participants occur. The number of possible scenarios and test cases is huge and renders a case by case approach, even for function prototyping and performance evaluation, almost impossible. In this work, a virtual development framework is proposed which allows performance testing under realistic traffic conditions by taking the interaction between SUT and other participants into account. A combination of a microscopic traffic simulation and a high-detailed vehicle simulation is utilized. To handle the interaction between both tools, a co-simulation framework with an interface layer for synchronization is developed which serves also as input for virtual sensors and prototype functions. The framework is demonstrated by a case study for a predictive adaptive cruise control (P-ACC) in a connected environment. This case study shows both the potential benefits of utilizing available information via new communication channels for ADAS and the applicability of the proposed framework.

H. Waschl (✉) · R. Schmied
Institute for Design and Control of Mechatronical Systems,
Johannes Kepler University, Linz, Altenberger Str. 69, 4040 Linz, Austria
e-mail: harald.waschl@jku.at

R. Schmied
e-mail: roman.schmied@jku.at

D. Reischl
LCM GmbH, Altenberger Str. 69, 4040 Linz, Austria
e-mail: daniel.reischl@lcm.at

M. Stolz
Virtual Vehicle Research Center, Inffeldgasse 21/A, 8010 Graz, Austria
e-mail: michael.stolz@v2c2.at

6.1 Introduction

In the last decade, the number of (advanced) driver assistance system (ADAS) available in production vehicles has been steadily increasing. At the present state, several assistance systems have been introduced which assist the driver in longitudinal (e.g., adaptive cruise control, autonomous emergency braking, or traffic jam assist) and lateral guidance (e.g., lane keeping assistant or blind spot warning) of the vehicle [3]. The assistance functions can be seen as iterative steps toward fully autonomous driving, where the system takes over control of the vehicle and fulfills the driving mission. ADAS can help to improve comfort, safety, and also fuel economy [16] and provide benefits for the driver. It is also expected that a positive impact on traffic throughput and network performance can be gained [9].

An important task is to evaluate the performance of such driving functions and assistance systems during development and certification. While for systems acting on a vehicle dynamics level, it is possible to test the individual functions and verify safety with a given number of test cases, for the advanced ADAS or even autonomous driving this task becomes much more complex [20]. These systems establish a perception of their environment under varying conditions, decide on a control, utilize the actuators, and interact with other traffic participants and infrastructure. Although the first task is already challenging, the interaction with other traffic participants and the number of potentially encountered scenarios can reach almost infinity.

To address this task, different approaches for testing and validation with varying levels of detail are proposed: Starting from component level tests, to simulation studies with software and hardware in the loop setups (S/HIL), up to tests with full prototype vehicles in real-world experiments. More details on the different levels and a survey on existing approaches can be found in [24]. In [17], a summary of the current methods and issues of testing is given and new, intelligence testing approach proposed. Zofka et al. [29] introduce a framework combining several components and their virtual and hardware representations. Co-simulation and modularity are a key aspect which allows to exchange components on SIL or HIL level. Other proposals suggest to use real vehicles on a closed road and deduce other traffic participants from a microscopic traffic simulation, see [11], where a real vehicle is transferred into a microscopic traffic simulation based on SUMO [14]. An approach, which uses real-world test drives including recorded maneuvers of other traffic participants to deduce scenarios and a high detail vehicle model including an ADAS prototype, is presented in [28].

To summarize, mixed virtual reality simulation and HIL-based approaches seem to be a promising path. The question remains how scenarios for such setups can be found. An novel method to generate interactive testing scenarios with multiple participants using game theory has been recently presented in [18]. For testing, it may be distinguished between safety-critical scenarios and performance relevant scenarios. Safety-critical events occur only seldom in real-world driving situations and need to be generated and tested explicitly. Therefore, methods for generation and accelerated testing are suggested; see [25, 27]. For performance testing (comfort, fuel economy), it is important to have realistic and typical behavior of other (human) traffic participants interacting with vehicle.

Within this work, we present a high-level virtual evaluation framework for ADAS and automated driving functions. The idea is to establish a co-simulation of multiple tools which can be extended to address special needs of the SUT. The main contributions are a new interfacing layer which is developed in MATLAB and responsible for synchronization and further serves as simulation master and a prototype predictive adaptive cruise control (P-ACC) which is used as test function. The interface layer establishes interfaces for virtual sensors and the ADAS prototype and handles the interaction between different tools. The focus is on urban city driving with moderate speeds, traffic lights, vehicle-to-vehicle or vehicle-to-infrastructure communication (V2X), and multiple participants. A prototype P-ACC is developed, which assumes V2X connections between all participants and infrastructure. It utilizes V2X information for a prediction of the future movement of preceding vehicles and considers them in a MPC-based control strategy to solve for the vehicles longitudinal control.

The intended application is on performance testing in varying and realistic environments, not on safety-critical scenarios. A potential advantage is that scenarios can evolve out of the traffic simulation. This is, to a certain extend, similar to the scenario design approach presented in [18]. In our approach, we rely on the traffic simulation tool to generate the surrounding traffic and consequently rely on the interaction models of the traffic simulation. A similar approach, using microscopic traffic and vehicle simulation, was presented recently in [13]. We extend the concept by introducing a dedicated interface layer, present a prototype ADAS as an exemplary case study, and discuss the potential benefits and properties of the proposed framework versus testing only with a single tool.

The rest of this chapter is organized as follows, in Sect. 6.2 the framework, interface layer, and used tools are introduced. Section 6.3 presents the development of the prediction models and control strategy for the predictive adaptive cruise control function using V2X and I2V communication, and the obtained results are shown in Sect. 6.4. Section 6.5 summarizes the presented approach and provides conclusions.

6.2 Virtual Development and Evaluation Framework for ADAS

The aim is to establish an evaluation framework for ADAS by combining two complementary simulation tools in a co-simulation and allow to develop and evaluate prototype ADAS functions in a realistic, virtual environment. It is assumed that only one vehicle, the so-called ego vehicle, is equipped with the prototype ADAS function which should be evaluated and "drives" through the simulated urban environment.

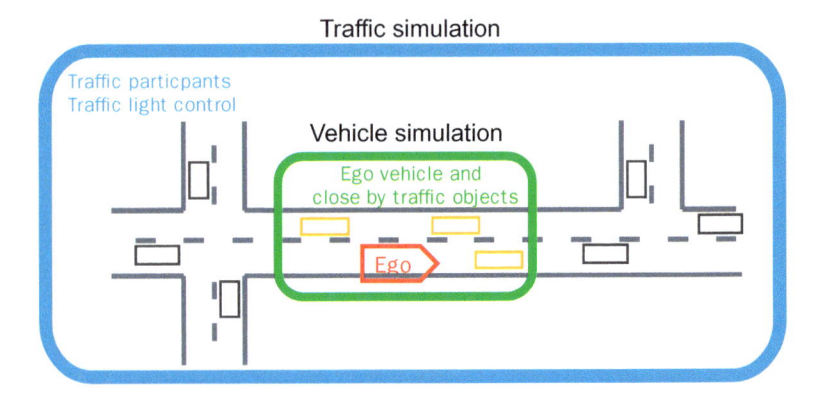

Fig. 6.1 Concept of co-simulation framework

6.2.1 Concept

The main concept can be imagined as a moving magnifying glass, where the simulation detail around the ego vehicle is increased and the other parts are operated in a lower resolution.

This approach allows to use the high fidelity and dynamic vehicle model of a vehicle simulation tool including available sensor models or driving functions, while the large-scale microscopic simulation provides the basic scenario and traffic participants movement and control. In other words, the microscopic traffic simulation serves as environment, including other passenger cars, public transport, infrastructure, or pedestrians.[1] By varying parameters in the simulation, such as signal timing or traffic composition or the ego starting coordinates and time, a large number of different use cases can be generated without the need to specify the trajectories and paths of all participants in the environment manually. This is one of the main advantages of the combination of both tools. The ego vehicle movement will be synchronized and represented as an external user-driven car which takes part in the traffic and interacts with the other participants. A sketch of this idea is presented in Fig. 6.1. Additionally, it should be noted that a high-level approach is pursued, where the main focus is on an object level and in the interaction between participants, infrastructure and ego vehicle equipped with the prototype ADAS function.

In the following, available simulation tools and the selected solutions for traffic and vehicle simulation will be discussed briefly. Different tools are described, however, the list is not considered to be complete. Unfortunately, it was not possible to perform a comparative study between different tools within this work; thus, only main properties and features of the ones used are described.

---

[1] In the presented example, only other passenger cars are considered as traffic participants. Depending on the used microscopic traffic simulation public transport or pedestrians can be considered too.

6.2.1.1 Microscopic Traffic Simulation—TraSim

For microscopic traffic simulation, many tools exist, such as Aimsun, SUMO, or PTV VISSIM [1]. The tools are mainly used by transportation researchers and engineers to evaluate and test traffic control strategies, public transportation concepts or road designs. In our case, PTV VISSIM is applied, a microscopic discrete traffic simulation [7]. One reason for the choice is that PTV VISSIM offers parameterized psychophysical driver models which are based on studies of human car-following and lateral vehicle guidance. It provides traffic models where participants are considered with varying stochastic parameters regarding their desired velocity, typical acceleration, and deceleration or following distance. For conflict areas, e.g., intersections or merging lanes, it features anticipated driving strategies. A heterogeneous traffic composition is modeled and even untypical behavior of participants such as peeking or overtaking on the same lane occurs, an advantage compared to other tools in which vehicles operate only on discrete lanes. The parameters are tuned to West European traffic and thus provide a reasonable initial setup.[2] Road geometry is defined by links, with possibly multiple lanes, and connectors between them defining the possible routes. In the proposed framework, the traffic control can be performed by PTV VISSIM, e.g., by offered signal controllers. Nonetheless, it is possible to integrate external components to provide the traffic signal control.

PTV VISSIM offers external interfaces to interact with the simulation and control dedicated vehicles in the network directly by the user. These interfaces are a *COM* interface and a *driving simulator* interface provided in form of a *dynamic-link library* (DLL) which can be accessed for example from MATLAB. The simulation itself is executed in discrete steps which can be triggered externally by the interfaces. This simplifies the use within a co-simulation framework, because a dedicated external master can be used.

6.2.1.2 Vehicle Simulation—VehSim

Similar to the microscopic traffic simulation, a large number of tools exists for vehicle simulation. A main difference compared to traffic simulation tools is the focus on a single vehicle (often called ego vehicle) and its environment which consists of the road and its properties. The environment can be extended by other traffic participants in the vicinity of the vehicle or external influences such as weather conditions. In these tools, the vehicle dynamics are more complex and detailed multibody models are typically used. This allows to cover the vehicle behavior in highly dynamic transient situations and to analyze the interaction between powertrain, suspension, tires, road, and chassis itself. Of course, the computational load is much higher than for a single vehicle in the microscopic traffic simulation.

---

[2]For the latter case study, the parameters of the car-following model were adapted to match the behavior recorded during real test drives.

In this work, IPG CarMaker [6] is used as vehicle simulation tool. It is a virtual test drive simulation, including a multibody vehicle model with suspension kinematics, tire models, variable road properties, complex configurable powertrain, and sensor models. It offers a high level of detail and provides interfaces to third-party tools, user-defined components, and even HIL setups for the full vehicles. Although the initial domain of this tool was focused on vehicle dynamics and control, recent updates improved the virtual environment, sensor, and traffic level. It is possible to implement other traffic participants as moving objects and control them externally. For visualization purposes, a virtual environment called IPG Movie is available which graphically represents the ego vehicle, the road, the environment, and other traffic participants. An example for the visualization of a co-simulation scenario in both tools respectively is depicted in Fig. 6.8. IPG CarMaker provides driver models for the ego vehicle, which can be parameterized, interact with other traffic objects and consider traffic signals. Lateral and longitudinal vehicle controls are split which becomes handy in case of ADAS function testing and allows to keep a lateral control active while longitudinal control is taken over by the prototype function. For function development and prototyping, IPG CarMaker provides a MATLAB-Simulink interface which allows to control the ego vehicle inputs or exchange components of the vehicle powertrain. Within this work, the gas and brake pedal of the vehicle will be used as main interface for the ADAS; alternatively, it is possible to directly interact on a powertrain, brake, or engine level. The interface further allows to control the other traffic participants and their movement, e.g., by defining their position or velocity profiles.

6.2.2 Environment Synchronization—EnSim Layer

To couple both tools, the interfacing layer *EnSim* is developed. The task of EnSim is to provide synchronization of all traffic participants, to address issues such as different sampling times and offer interfaces for virtual sensors and prototype functions. It is implemented as a custom code Simulink block in MATLAB-Simulink within the IPG CarMaker for Simulink environment. The EnSim layer serves as master for the whole simulation, respectively experiment. It handles the information flow between tools, coordinate transformations and provides interfaces for virtual sensors and prototype functions. Furthermore, it triggers the discrete simulation steps in the traffic simulation (TraSim) environment.

A schematic representation is given in Fig. 6.2. In most cases, the TraSim is operating with a different and larger discrete step time (typically in the range of 100 ms), whereas the vehicle simulation (VehSim) operates with 1 ms. Consequently, interpolation and extrapolation to synchronize the traffic objects are necessary. The exchange of information on a tool level, i.e. between simulation environments, is assumed to be instantaneous and ubiquitous. Tools share all information which necessary for their operation, as well as visualization and evaluation. Effects, such as transport delays or sensor errors, are only considered within the driving functions or sensor models.

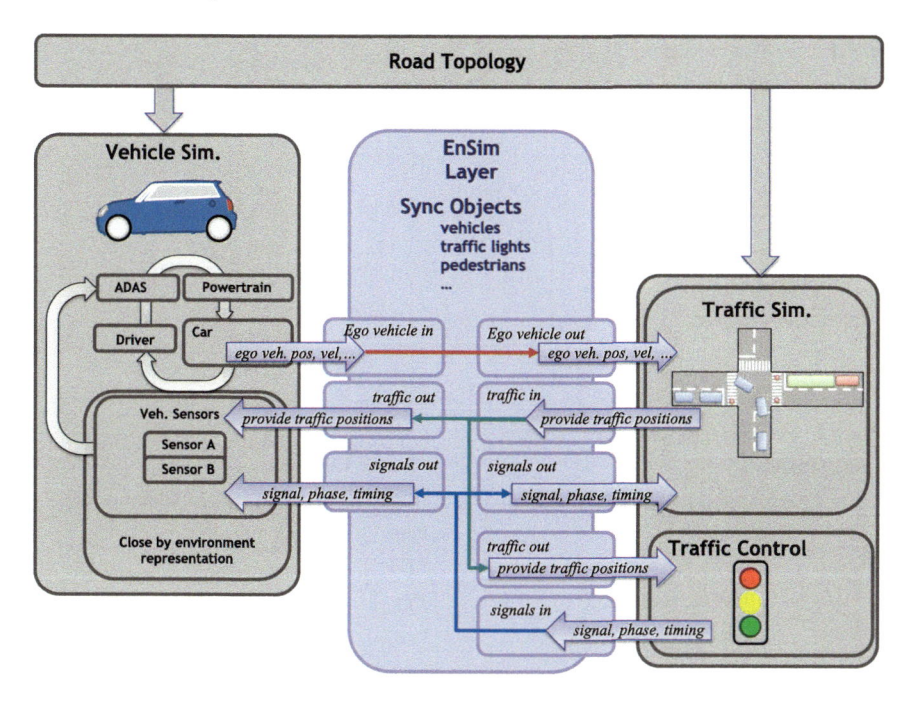

Fig. 6.2 Interaction scheme of development and evaluation framework

The VehSim serves as master for the position and movement of the ego vehicle in the road network. The TraSim receives the ego movement and provides the current information on other traffic participants and traffic lights to the VehSim, where close by object, movements are updated accordingly. The ADAS operating inside in the ego vehicle, i.e., the control SUT, receives the information from the EnSim layer by virtual sensors which can represent radar sensors, cameras, or on-board units (OBU) in case of V2X communication. These virtual sensors can be seen as filters for the full information available in the tool level and provide limited information to the control function.

Additionally, a synchronization of the traffic objects in the VehSim is performed, which allows the interaction of driver or inbuilt sensor models in the VehSim with the current traffic situation.

6.2.2.1 Interfaces

In both tools, the identical 3D Cartesian coordinate system x, y, z is utilized to describe the absolute position of the participants.[3] It is assumed that the vehicle

---

[3]For location-based control strategies, a mapping to global GPS coordinates can be performed easily. However, even in this case on a tool level, it can be beneficial to use an identical coordinate system.

Fig. 6.3 Coordinates used for synchronization between tools

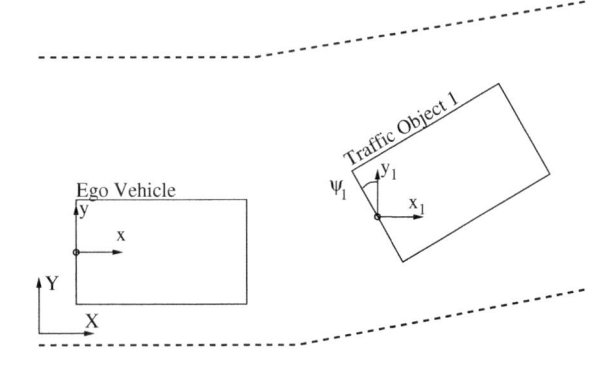

Table 6.1 Interface lists for data exchange

Traffic Objects for $j = 1 \ldots N_{V,TraSim}$ vehicles		Traffic Lights for $m = 1 \ldots N_{TL,TraSim}$ traffic lights	
Item	Comment	Item	Comment
ID_j	–	$ID_{TL,m}$	–
P_j	Position traffic object j	$P_{TL,m}$	Position of traffic light m
v_j	Velocity of traffic object j	$\sigma_{TL,m}$	Current signal state
Δx_j	Distance to ego in x	$t_{\sigma TL,m}$	time to next signal change
Δy_j	Distance to ego in y		
$Stat_j$	Flag to set visibility in VehSim		

follows the road without vertical displacement, and thus the z position is omitted. The position P_j of each traffic participant j is then defined by the x, y coordinates and the yaw angle, $P_j = [x_j, y_j, \psi_j]$, additionally the velocity information is described by $v_j = [v_x, v_y]$. In case of the ego vehicle, only P and v are used without a subindex. For each vehicle, a specific coordinate system, as depicted in Fig. 6.3, is introduced. The origin of this system is at the center of the rear bumper. This choice is motivated by the provided coordinates of the used tools.

One remark regarding the setup is that both tools need to share the same road topology, at least for the desired route of the ego vehicle. This requires an identical representation in both tools. In the current framework, the PTV VISSIM road network serves as master definition which is then mapped to the road definition in IPG CarMaker. The desired ego vehicle trajectory is parsed in PTV VISSIM and then imported after some post processing as described for real-world test drives in [28].

To synchronize the elements in EnSim, a common object definition is set up for traffic objects and traffic lights.[4] The items are listed in Table 6.1 and are used for environment synchronization in the VehSim and virtual sensors for the ADAS, respectively. Some partially redundant information is stored, e.g., absolute and

---

[4]In this case, a simplified position-oriented representation is used. In more complex intersections, additionally signal groups and routes would need to be defined and synchronized.

relative position to the ego vehicle, where the first one is used in the environment synchronization and the latter one for the virtual sensors. The current signal of a traffic light is defined by $\sigma_{TL,m}$, where only between $stop \rightarrow 1$ and $go \rightarrow 0$ is distinguished.

6.2.2.2 Implementation

As mentioned above, the EnSim layer is implemented as custom MATLAB-Simulink block. The components and information flow directions are depicted in Fig. 6.4.

The tasks and the basic procedure are described in the pseudo-code listed in Algorithm 1.

In the following, the functions of EnSim and the tool interfaces of TraSim and VehSim are briefly described:

readEgoState: Obtains in each simulation step, the current ego vehicle position and speed from the VehSim. Therefore, VehSim offers runtime dictionary variables, which are directly available during in MATLAB-Simulink as custom input blocks; see [6].

SetEgo: Sets the ego vehicle position and speed in the TraSim by accessing the driving simulator interface of PTV VISSIM from MATLAB via a function call; see [22].

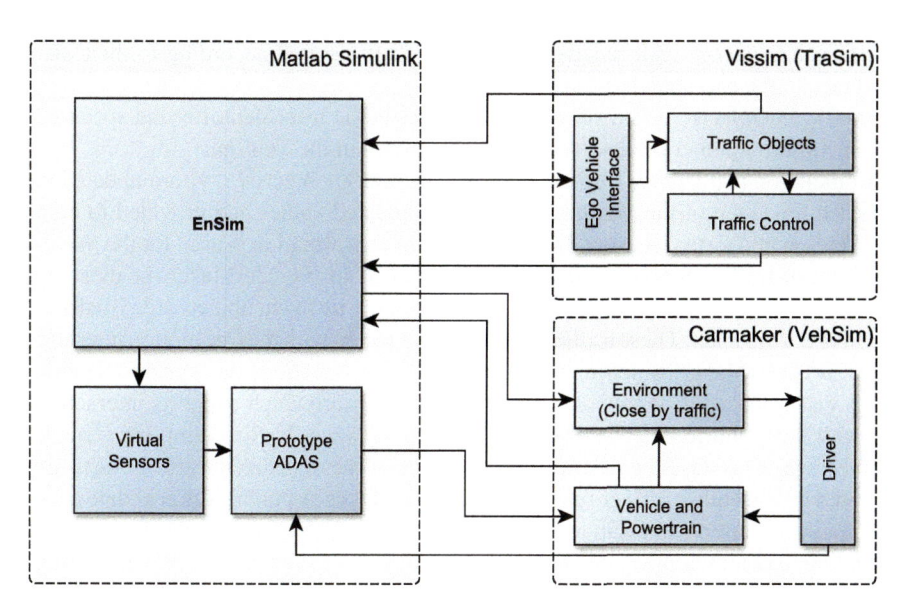

Fig. 6.4 Components and information flow of the framework. EnSim is implemented in MATLAB-Simulink and serves as simulation master process

Algorithm 1 EnSim

1: Initialize $t_{sim} = 0$, load and setup road network and experiment;
2: **while** Simulation running **do**
3:     $[P, v] = readEgoState(VehSim)$
4:     **if** mod($t_{sim}, T_{s,TraSim}$)=0 **then**   ▷ Synchronize and trigger TraSim step
5:        $TraSim \leftarrow SetEgo(P, v)$
6:        $[ID, P, v] = GetTrafficObjects(TraSim)$
7:        $[ID_{TL}, P_{TL}, \sigma_{TL}, t_{\sigma,TL}] = GetTrafficLights(TraSim)$
8:        $TraSim \leftarrow TriggerSimStep()$
9:        $SortTrafficObjects()$ ▷ Sort according to distance to ego
10:       $FilterTrafficObjects()$ ▷ Filter close by objects to ego
11:       $[VehSim, VirtSens] \leftarrow UpdateTrafficLights()$
12:     **end if**
          ▷ Calculate updated object and traffic light information w.r.t. P movement and extrapolation
13:     $[VehSim, VirtSens] \leftarrow UpdateTrafficObjects()$
14:     $P_j^+ = P_j^- + v_j \cdot T_{s,VehSim} \forall j \in N_{V,TraSim}$ ▷ Extrapolate traffic movement
15:     $t_{sim} = t_{sim} + T_{s,VehSim}$ ▷ Increment time with T_s of VehSim
16: **end while**

GetTrafficObjects: Extract all traffic objects currently existing in TraSim. For each vehicle, a position, velocity, and ID are received and stored.

GetTrafficLights: Extract the current signal state of all traffic lights in the network and calculate the future signal switching times.

TriggerSimStep: Execute a single step in the TraSim environment.

SortTrafficObjects: All received traffic objects are sorted according to their distance to the ego vehicle.

FilterTrafficObjects: The sorted traffic object list is truncated after the specified maximum number of observed close by vehicles in the VehSim.

Extrapolation and *UpdateTrafficObjects*: The traffic objects P_j, v_j are updated in each time step of the VehSim by their extrapolated values and provided to both VehSim and virtual sensors. In the VehSim, this information is used for the movement of traffic objects. As described in Fig. 6.1, in IPG CarMaker the close by traffic is represented by traffic objects. To this end, up to ten objects are visualized in IPG CarMaker. These traffic objects are controlled with the free motion interface of IPG CarMaker by setting position and yaw. A benefit of this synchronization is that it makes it possible to use the driver function which supports interaction with these traffic objects. The extrapolation between TraSim sampling steps is necessary, because the movements and states of the visualized traffic objects are used by the built-in driver function in IPG CarMaker. A pure position update each TraSim sampling step can lead to undesired driver behavior. During the extrapolation, a potential mismatch between TraSim and VehSim can occur, e.g., during sharp turns. This mismatch is corrected in the next sampling instant of the TraSim, once new information is available.

UpdateTrafficLights: Similar to the traffic objects, the signal state of all traffic lights is visualized in VehSim to, so the driving function may react accordingly. Additionally, this information is provided to the virtual sensors and so available for the control development.

EnSim is executed with the same discrete sampling time as VehSim. For the selected VehSim tool, IPG CarMaker for Simulink, the sampling time is fixed to $T_{s,VehSim} = 1$ ms. All tools are running as instances on the same machine. Alternatively, it is possible to use multiple machines; in this case, the interfaces need to be replaced by network connections. Besides the driver functionality, the synchronization makes it possible to use vehicle sensors which are available in IPG CarMaker. Another benefit of the synchronization is opportunity to use the IPG Movie interface in IPG CarMaker for visualization of the traffic and signals; see Fig. 6.8 or the presentation interface available in PTV VISSIM.

Virtual sensors

The control algorithm for the ADAS functions requires in most cases its own environment representation. Within the framework, this step is performed by virtual sensors which provide the algorithm with identical information as a "real" sensor would do. Different to the synchronization on an environment level, where full information is shared, these sensors can be seen as a synchronization on a function level. The implemented interfaces are similar to the ones of real sensors installed in a vehicle, e.g., object lists provided by a radar sensor. This allows to streamline the transfer to a real vehicle setup because only the interfaces have to be adapted but the same information is already provided. Typically, the interface is based on object lists which are described in Table 6.1 and the virtual sensors can essentially be seen as filters for the list entries.

As mentioned above, the main purpose of the proposed framework is on the functional testing of the control algorithm and its interaction with the environment. Thus, sensor stimulation or effects on a physical sensor level are disregarded and high-level approach is followed. Still, the virtual sensor can include effects such as jitter or occlusion.

6.3 Case Study Predictive Adaptive Cruise Control utilizing V2V and I2V Communication

The presented framework will be applied in a case study of a prototype ADAS function. In this example, a predictive adaptive cruise control (P-ACC) is considered which utilizes information of surrounding traffic participants and traffic lights transmitted by its communication.

P-ACC is an adaptive cruise control (ACC) strategy which exploits an estimation of the future behavior of surrounding traffic participants to improve safety, economy, and comfort. A cooperative urban environment where it is assumed that vehicles and traffic lights transmit information about their state constitutes the basis for the approach. The objective is to control the longitudinal dynamics of the ego vehicle at its driver's velocity setpoint while ensuring collision avoidance with other vehicles and maximizing economy and driving comfort. The case study investigates a model predictive control (MPC) approach in combination with an online prediction of the leading vehicle's driving behavior. The prediction model relies on the cooperative information from surrounding objects.

The virtual sensor module simulates available V2X communication via an onboard unit (OBU). The communication layer and protocols are simplified[5], and the information from all vehicles and traffic lights within sensor range is assumed to be available. It should be mentioned that an extension to the full communication interface and messages would be possible due to the modular structure of the framework. The same is true for other sensor models, such as radar sensors. In this example, the virtual sensor model can be extended by a virtual radar cone for the detection range and consider occlusion effects. Most of these extensions can be done on the object list level and do not require changes in the framework itself.

6.3.1 Control Problem Description

In the P-ACC approach, we consider a single-lane road in a dense urban traffic scenario. We assume that information about surrounding vehicles and traffic light states is available for the ego vehicle control due to V2X communication modules; see Fig. 6.5. Specifically, it is assumed that each vehicle j transmits its actual position P_j and velocity v_j and each traffic light sends its position P_{TL}, signal phasing σ_{TL}, and time to the next signal change T_σ. The ego vehicle receives data within the vehicle-to-vehicle (V2V) and infrastructure-to-vehicle (I2V) communication range $d_{max,V2V}$ and $d_{max,I2V}$ only and processes this data via an OBU. For the processing, the virtual sensor of the framework is used. It should be mentioned that in this work communication delays or losses are neglected since we rather focus on the ego vehicle's control approach and the methodology than on the communication processes and data transmission. In a real application, e.g., on a HIL setup in a test vehicle, an encapsulation layer is necessary, which provides the information in the specified format to the P-ACC.

The task to be solved in this work is to develop a P-ACC strategy following different objectives while providing vehicle safety in terms of avoiding rear-end collision with leading vehicles. Further, additional environmental influences and disturbances should be considered within the control. We can summarize the control objectives in the following:

---

[5]The communication layer and formats defined of ETSI for CAM, SPAT, and MAP, see [8], are not implemented.

6 A Virtual Development and Evaluation Framework for ADAS

Fig. 6.5 Adaptive cruise control within a cooperative environment

- Provide an economic and comfortable driving behavior;
- Follow the drivers velocity setpoint v_{set};
- Ensure collision avoidance;
- Provide vehicle-following behavior which is acceptable for human passengers;
- Consider traffic light signals.

For the ego vehicle's control design, we will follow a model-based control approach. For this reason, a compromise between model accuracy and low level of model complexity should be found. Concerning the ego vehicle model, we suggest a simplified model for the longitudinal dynamics as proposed also in [23]. In the model, d denotes the distance to the leading vehicle (index L) driving with velocity v_L, v and a denote the ego vehicle's velocity and acceleration, respectively. Linear friction is considered by the friction coefficient c_f and acceleration dynamics via a first-order system with time constant τ_p. The pedal gain c_p describes the dynamics from the pedal input $u(t)$ to the vehicle acceleration $a(t)$.

$$\begin{aligned}\dot{d}(t) &= v_L(t) - v(t) \\ \dot{v}(t) &= a(t) \\ \dot{a}(t) &= -\frac{c_f}{\tau_p}v(t) - \left(\frac{1}{\tau_p} + c_f\right)a(t) + \frac{c_p}{\tau_p}u(t).\end{aligned} \quad (6.1)$$

The unknown parameters c_f, τ_p, and c_p are identified based on real-world and simulation driving data using the high-fidelity VehSim environment IPG CarMaker. Different driving cycles representing urban traffic scenarios within a velocity range of 0–16 ms^{-1} are considered as identification data. The resulting parameter values are given in Table 6.2.

The goals mentioned above, namely developing a comfort- and economy-oriented P-ACC approach can be described by an optimal control problem. Besides that, system constraints have to be considered on the one hand to assure safety by keeping a

Table 6.2 Vehicle model parameters

Parameter	c_f	c_p	τ_p
Value	0.0253/s	3.3986/s²	0.1278/s

safety distance to the leading vehicle and on the other hand to account for environmental and vehicle limits. The safety distance is chosen according to the well-known constant time headway policy

$$d(t) \geq d_{\min} + t_h v(t) \tag{6.2}$$

with the minimum distance at standstill $d_{\min}$ and the constant time headway t_h. Additionally, speed limit and input constraints should be included

$$0 \leq v(t) \leq v_{\lim} \tag{6.3}$$
$$u_{\min} \leq u(t) \leq u_{\max}. \tag{6.4}$$

Also, red traffic lights have to be considered by the controller, i.e., the distance to the upcoming traffic light $d_{\text{TL}}(t)$ must be positive in case of a red signal phase $\sigma_{\text{TL}}(t) = 1$

$$d_{\text{TL}}(t) \geq 0 \quad \text{if} \quad \sigma_{\text{TL}}(t) = 1. \tag{6.5}$$

Generally, we can describe the overall problem with a constrained optimal control problem introducing the objective function $J(x(t), u(t))$ depending on the system states $x(t) = [d(t), v(t), a(t)]$ and the input $u(t)$ and summarizing the system dynamics (6.1) and constraints (6.2), (6.3), (6.5) in the state set $X(\sigma_{TL})$ and input set U. The admissible state set $X(\sigma_{TL})$ depends on the signal state, which is necessary to consider red traffic lights which cannot be passed. Additionally, the disturbance $w(t) = v_L(t)$, i.e., the velocity of the leading vehicle is restricted by the set W.

$$\min_{u(t)} J(x(t), u(t)) \tag{6.6a}$$

$$\text{s.t.} \quad \dot{x}(t) = \bar{A}x(t) + \bar{B}_u u(t) + \bar{B}_w w(t)$$
$$x(t) \in X(\sigma_{\text{TL}})$$
$$u(t) \in U \tag{6.6b}$$
$$w(t) \in W$$

To solve this optimal control problem (OCP), the disturbance trajectory $w(t)$ has to been known in advance. Since this is impossible for online applications in everyday traffic situations, a prediction model to estimate the expected disturbance profile has to be investigated.

6.3.2 V2X Based Traffic Prediction

The information on surrounding vehicles and traffic lights provided to the ego vehicle by the OBU can be used to investigate a prediction model which estimates the future driving trajectories of leading vehicles. Having such a prediction available enables to solve the OCP (6.6) which represents the key element to realize anticipatory ego vehicle control. The prediction model in this work differs essentially from other approaches; see e.g. [16, 21]. Instead of predicting only the direct leader, the entire string of vehicles within the communication range is considered. Since we rely on dense urban traffic situations, we focus on the analysis and modeling of human car-following behavior. Once a model for this is established, an entire string of vehicles can be modeled by simply augmenting the model where one vehicle provides the input for its follower.

Modeling of human vehicle-following behavior has been studied since many years; see [5] for a review. Different methods have been proposed for implementing realistic vehicle-following behavior in traffic simulators, e.g., [4, 10, 15, 26]. The modeling approach here relies on a linear time-delayed model structure and similar to an approach first proposed by [12].

It is assumed that in a vehicle-following scenario, the acceleration of vehicle j can be approximated by

$$a_j(t) = k_\mathrm{d} \left(d_j(t - T_\mathrm{r}) - t_\mathrm{h} v_j(t - T_\mathrm{r}) - d_0\right) \\ + k_\mathrm{v} \left(v_{j-1}(t - T_\mathrm{r}) - v_j(t - T_\mathrm{r})\right) \tag{6.7}$$

Hence, each vehicle in the string is modeled as a time-delayed PD controller with control gains k_d, k_v tracking a time headway distance $d_0 + t_\mathrm{h} v_j(t)$ and relative velocity with a reaction time T_r.

To determine the unknown parameters $\rho = \begin{bmatrix} k_\mathrm{d} & k_\mathrm{v} & t_\mathrm{h} & T_\mathrm{r} & d_0 \end{bmatrix}$ of the car-following model, we use driving data from the traffic simulation tool PTV VISSIM. The car-following model there is an extension of the *Wiedemann 74* model and extensively trained with real-world driving data. For each car, a different parameter set, defining its driving behavior is assigned in PTV VISSIM. This leads to a heterogeneous traffic, representing different driving behaviors as they appear in real-world traffic. For the prediction model, we identify a set of parameters ρ describing the average driving behavior, i.e., for prediction of the preceding cars, we assume identical (average) car-following behavior. The prediction model parameters are determined by a numerical least squares routine to minimize the error between model and PTV VISSIM driving data. Figure 6.6 illustrates the modeled and measured distance and velocity profile for an exemplary simulation driving cycle taken from PTV VISSIM data.

To use the introduced car-following model (6.7) within the control formulation, two aspects need to be considered: First, a model for the foremost vehicle within the V2V communication range needs to be established since (6.7) requires the velocity of the leader as input which is not available for the foremost vehicle. Second, the model has to be converted into a suitable prediction form with the velocity of the ego vehicle's leader $w(t) = v_\mathrm{L}(t)$ as model output, cf. (6.6).

The first issue is addressed by assuming that the foremost vehicle tends to drive at the velocity limit v_lim and stops at red traffic lights. Hence, we can identify a parameters set $\rho_1(\sigma_\mathrm{TL}(t))$ depending on the actual traffic light signal. If there is a red traffic light ahead, i.e., $\sigma_\mathrm{TL}(t) = 1$ we assume the traffic light to be a standing

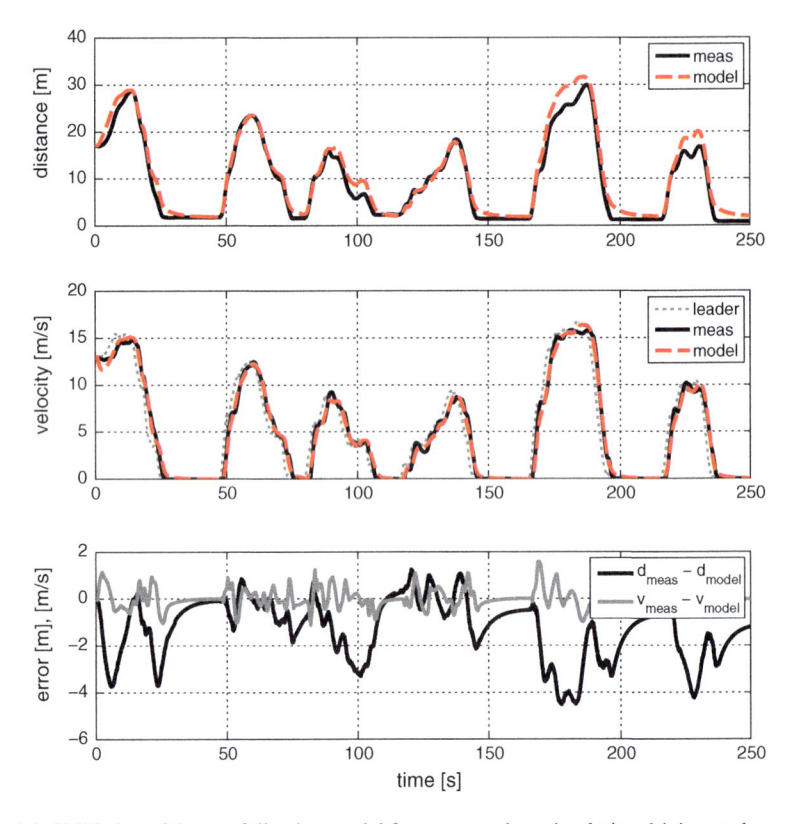

Fig. 6.6 Validation of the car-following model for an exemplary simulation driving cycle

object at position p_{TL}. In this way, the traffic light can be considered as the foremost vehicle's leader using the car-following model (6.7).

We can now address the second aspect and establish a compact prediction model considering all vehicles within the communication range. To do so, we introduce the state $\xi_j(t) = \begin{bmatrix} d_j(t) & v_j(t) \end{bmatrix}^T$ for each vehicle in the platoon. Defining $v_{j-1}(t) = \begin{bmatrix} 0 & 1 \end{bmatrix} \xi_{j-1}$ as the input to the model of the jth driver a linear time-delayed state-space representation can be found

$$\dot{\xi}_j(t) = \begin{bmatrix} 0 & -1 \\ 0 & 0 \end{bmatrix} \xi_j(t) + \begin{bmatrix} 0 & 0 \\ k_d & -k_d t_h - k_v \end{bmatrix} \xi_j(t - T_r)$$
$$+ \begin{bmatrix} 0 & 1 \\ 0 & 0 \end{bmatrix} \xi_{j-1}(t) + \begin{bmatrix} 0 & 0 \\ 0 & k_v \end{bmatrix} \xi_{j-1}(t - T_r) + \begin{bmatrix} 0 \\ -k_d d_0 \end{bmatrix} \quad (6.8)$$
$$\xi_j(0) = \begin{bmatrix} d_j(0) & v_j(0) \end{bmatrix}^T .$$

Since the P-ACC controller will be implemented as discrete time model predictive controller (MPC), discretization of (6.8) using the sampling time T_s is necessary resulting in

$$\xi_{j,k+1} = F_1 \xi_{j,k} + F_2 \xi_{j,k-K_r} \\ + G_1 \xi_{j-1,k} + G_2 \xi_{j-1,k-K_r} + \beta \quad (6.9)$$

with $T_r = K_r T_s$.

Introducing the traffic light state $\xi_{0,k}(\sigma_{TL})$ as input for the foremost vehicle, we can then put (6.9) into augmented state-space form to describe the overall vehicle platoon behavior within a single model.

Defining the augmented state vector $\mathbf{z}_k = \begin{bmatrix} \xi_{0,k} & \xi_{1,k} & \ldots \end{bmatrix}^T$, the model can be stated as

$$\mathbf{z}_{k+1} = \mathbf{F}_1 \mathbf{z}_k + \mathbf{F}_2 \mathbf{z}_{k-K_r} + \boldsymbol{\beta} \quad (6.10)$$

$$\mathbf{F}_l = \begin{bmatrix} 0 & 0 & 0 & \cdots & 0 \\ G_j & F_j & 0 & & \\ 0 & G_j & F_j & & \vdots \\ \vdots & & \ddots & \ddots & \\ 0 & \cdots & & G_j & F_j \end{bmatrix}, \quad l \in \{1, 2\}$$

$$\boldsymbol{\beta} = \begin{bmatrix} \beta & \cdots & \beta \end{bmatrix}^T$$

The velocity $v_{L,k}$ of the ego vehicle's direct leader appears in the very last element of the state $\mathbf{z}_k$ and is hence given by

$$v_{L,k} = \begin{bmatrix} 0 & \cdots & 0 & 1 \end{bmatrix} \mathbf{z}_k . \quad (6.11)$$

6.3.3 P-ACC Design

To target the objectives stated in Sect. 6.3.1 using the optimal control problem formulation (6.6), the following objective function is suggested

$$J(x_k, u_k) = \sum_{k=0}^{N} r_v (v_k - v_{\text{set}})^2 + r_u u_k^2 + r_{\Delta u} \Delta u_k^2 \quad (6.12)$$

penalizing deviations from the drivers velocity setpoint v_{set} as well as absolute values and changes in the pedal position u_k. To make the control approach suitable for online application, a model predictive control strategy is investigated solving the OCP in every sampling step k across a prediction horizon N_{PH} and using feedback information on the actual states x_k and disturbances $w_k = v_{L,k}$. Incorporating the driver prediction model (6.10) into the MPC framework results in a predicted velocity profile for the leading vehicle(s).

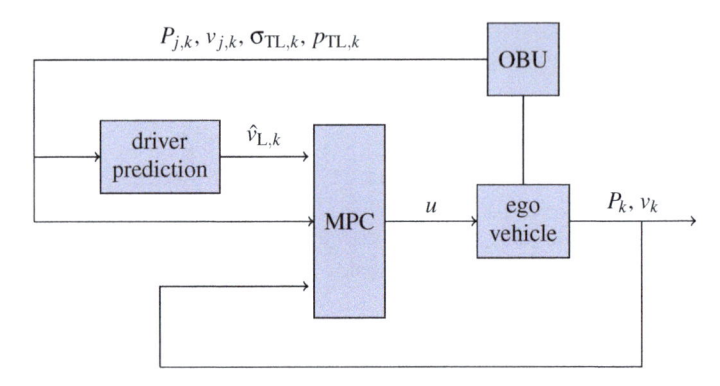

Fig. 6.7 Block diagram of control structure: The driver prediction model uses position and velocity information from the OBU of all vehicles $j \in \{1, \ldots, N_{\text{veh}}\}$ within the communication range to estimate the future velocity profile $\hat{v}_{L,k}$ of the ego vehicle's direct leading car

The proposed control structure is depicted in Fig. 6.7. The prediction model receives the data from the V2X communication module of the ego vehicle and estimates the prospective velocity profile $\hat{v}_{L,k}$ of the leader across the prediction horizon N_{PH} of the MPC according to (6.10), (6.11) at each time instant k. The inputs of the MPC besides $\hat{v}_{L,k}$ are traffic light information and the actual ego vehicle states p_k, v_k.

To make the OCP (6.6), (6.12) suitable for an MPC framework, some aspects need to be taken into account: The condition in the traffic light constraint (6.5) is realized using the so-called Big-M method which leads to a mixed integer (MI) program [2]. The cost function in (6.12) is of quadratic form which hence would result in a MI quadratic program (MIQP). To reduce the computational complexity and make use of efficient solvers, the problem is reformulated as MI linear program (MILP). This is done by approximating quadratic functions by the maximum of a set of N_f piecewise affine functions

$$\zeta(\alpha) = \alpha^2 \approx \max\left(\Theta \begin{bmatrix} \alpha \\ 1 \end{bmatrix}\right), \quad \Theta = \begin{bmatrix} \theta_{1,1} & \theta_{1,0} \\ \vdots & \vdots \\ \theta_{N_f,1} & \theta_{N_f,0} \end{bmatrix}. \quad (6.13)$$

A quadratic cost may be approximated by a linear one in the following way:

$$\min_{\alpha} \alpha^2 \approx \min_{l} l, \quad \text{s.t.} \quad l\mathbf{1} \geq \Theta \begin{bmatrix} \alpha \\ 1 \end{bmatrix} \quad (6.14)$$

To write the OCP in receding horizon formulation with prediction horizon N_{PH}, we use the notation $\hat{x}_{i|k}$ to denote the i step ahead prediction of variable x and time instant k. The decision variables are summarized with $\mathbf{u}_k = \hat{u}_{i|k}$, $\mathbf{l}_k = \begin{bmatrix} l_{u,i|k} & l_{\Delta u,i|k} & l_{\Delta v,i|k} \end{bmatrix}$, $\forall i \in \{0, \ldots, N_{\text{PH}} - 1\}$, where $\mathbf{l}_k$ arises from the piecewise affine approximation (6.14) of the cost terms in (6.12) and weighting vector $\mathbf{r} = \begin{bmatrix} r_u & r_{\Delta u} & r_v \end{bmatrix}$.

$$\min_{\mathbf{u}_k, \mathbf{l}_k} \sum_{i=0}^{N_{\text{PH}}-1} \mathbf{r}\, \mathbf{l}_{i|k} \tag{6.15}$$

s.t.
$$\begin{aligned}
\hat{x}_{i+1|k} &= A\hat{x}_{i|k} + B_u \hat{u}_{i|k} + B_w \hat{w}_{i|k}, \quad \hat{x}_{k|0} = x_k \\
\hat{z}_{i+1|k} &= \mathbf{F}_1 \hat{z}_{i|k} + \mathbf{F}_2 \hat{z}_{i-K_r|k} + \beta \\
\hat{w}_{i|k} &= \begin{bmatrix} 0 & \dots & 0 & 1 \end{bmatrix} \hat{z}_{i|k} \\
\mathbf{l}_{i|k} \mathbf{1} &\geq \Theta\, \tilde{\mathbf{u}}_{i|k}, \\
\hat{x}_{i|k} &\in X(\sigma_{\text{TL}} i | k) \\
\hat{u}_{i|k} &\in U
\end{aligned} \tag{6.16}$$

where

$$\Theta = \begin{bmatrix} \Theta_u & \Theta_{\Delta u} & \Theta_{\Delta v} \end{bmatrix}^T, \quad \Theta_j = \begin{bmatrix} \theta_{j,11} & \theta_{j,10} \\ \vdots & \vdots \\ \theta_{j,N_j 1} & \theta_{j,N_j 1} \end{bmatrix}, \quad j \in \{u, \Delta u, \Delta v\}$$

$$\tilde{\mathbf{u}}_{i|k} = \begin{bmatrix} \hat{u}_{i|k} & 1 & \Delta \hat{u}_{i|k} & 1 & \Delta \hat{v}_{i|k} & 1 \end{bmatrix}, \quad \Delta \hat{u}_{i|k} = \hat{u}_{i|k} - \hat{u}_{i-1|k}, \quad \Delta \hat{v}_{i|k} = \hat{v}_{i|k} - v_{\text{set}}$$

and A, B_u, B_w result from discretizing the vehicle dynamics (6.1).

6.4 Case Study—Performance Evaluation

6.4.1 Route and Setup

To demonstrate the virtual evaluation environment, a simple demo example is set up consisting of a straight road and a series of three traffic lights.

The P-ACC takes over the longitudinal control of the ego vehicle and is implemented in the MATLAB-Simulink extension of IPG CarMaker. The lateral control of the ego is performed by the available driver function. All other traffic participants and the traffic light timing are controlled by PTV VISSIM. The traffic lights operate with a fixed timing and the information of the signal timing and other traffic participants is available to the ADAS via the virtual OBU sensor.

Figure 6.8 shows an exemplary screenshot of the simulation where the graphical user interfaces of PTV VISSIM and IPG CarMaker are presented at the top and bottom, respectively. The arrows illustrate the co-simulation, i.e., the synchronization of both simulation platforms by the EnSim layer.

For the road, a speed limit of $v_{\text{lim}} = 50$ km/h is assumed and overtaking of vehicles is prohibited for the ego vehicle as well as for all other traffic participants.

The MPC problem (6.15), (6.16) introduced above is implemented in MATLAB using the YALMIP toolbox [19] with the solver GUROBI to solve the MILP problem. The simulation parameters are presented in Table 6.3.

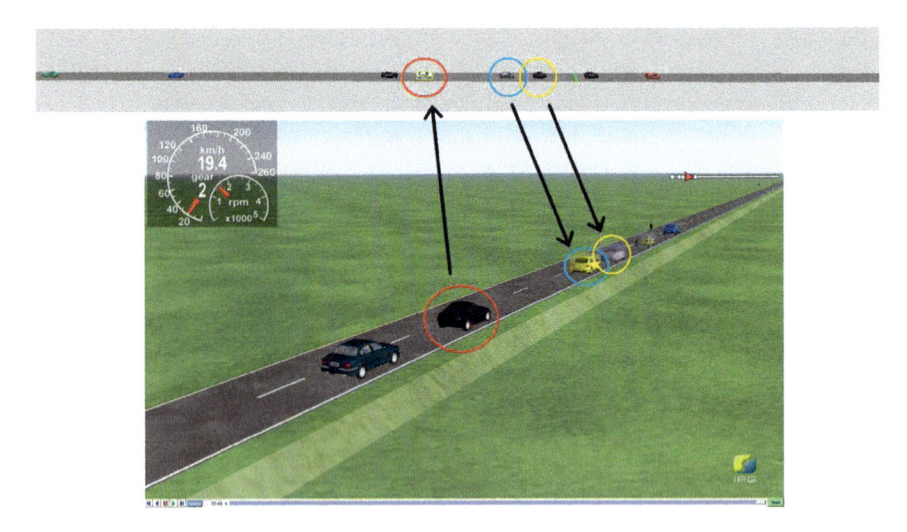

Fig. 6.8 Screenshot of simulation tools PTV VISSIM and IPG CarMaker used for co-simulation

Table 6.3 Simulation and control parameters

Description	Variables	Values
Simulation parameters	T_s, N_{PH}	0.2 s, 50
Time headway	t_h	1 s
MPC weight. par.	r_u, $r_{\Delta u}$, r_v	5, 20, 0.2
V2X parameters	$d_{max,V2V}$, $d_{max,I2V}$	200, 400 m

6.4.2 Results

Figure 6.9 depicts results of the P-ACC approach evaluated with the virtual co-simulation framework. The graphs show position and velocity trajectories of the ego vehicle as well as of leading traffic participants. Additionally, the traffic light signal phasing are illustrated with orange and red lines accordingly. From Fig. 6.9, it can be seen that first the ego vehicle's velocity profile shows a smooth and comfortable trajectory compared to the leading vehicles. Second, at $t \approx 38$ s the ego vehicle realizes that passing the upcoming traffic light at a green phase will not be possible and therefore starts braking even before the traffic light actually switches to red. Similarly, since the ego vehicle knows in advance when the traffic light will switch to green again, it starts accelerating during the end of the red phase to pass the traffic light just at the moment it switches to green. In this way, it is possible to preserve kinetic energy which reduces fuel consumption and engine emissions.

To further highlight the advantages of the V2X-based prediction model within the P-ACC, the approach is compared to a P-ACC with constant disturbance prediction, i.e., all vehicles are assumed to drive with constant velocity for the entire prediction horizon N_{PH}. Figure 6.10 presents a comparison of the state trajectories and the pedal input u_k. Due to the V2X-based prediction, the future velocity profile of the

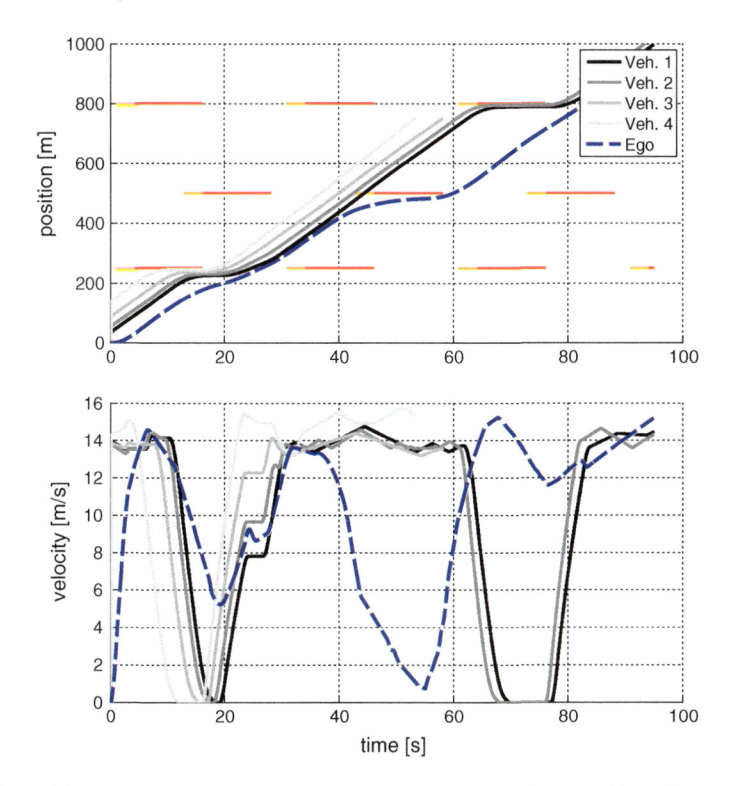

Fig. 6.9 Position and velocity trajectories of leading vehicles and the ego vehicle. Orange and red lines denote the according traffic light states

leading vehicle is estimated much more precisely, leading to higher comfort and smoother trajectories. Knowledge of the traffic light signal phasing in advance allows an accurate prediction of braking and acceleration maneuvers of leading vehicles and consequently an adaptation of the ego vehicle's speed according to these maneuvers.

The advantage of the V2X-based P-ACC approach becomes clearly visible when comparing the fuel consumption for the considered test scenario: While P-ACC with constant disturbance prediction leads to a total fuel consumption of 80.42 g, the P-ACC with V2X-based prediction model results in a total consumption of 64.62 g which equals a reduction of fuel consumption of 20%.

The presented prediction model for the car-following behavior uses the deterministic average and does not take into account uncertainties in the car-following behavior which will be addressed in future work. The presented P-ACC approach is sensitive to prediction accuracy, i.e., wrong predictions can lead to constraint violations and consequently performance losses. To address this issue, it is planned to incorporate the stochasticity in the car-following prediction model.

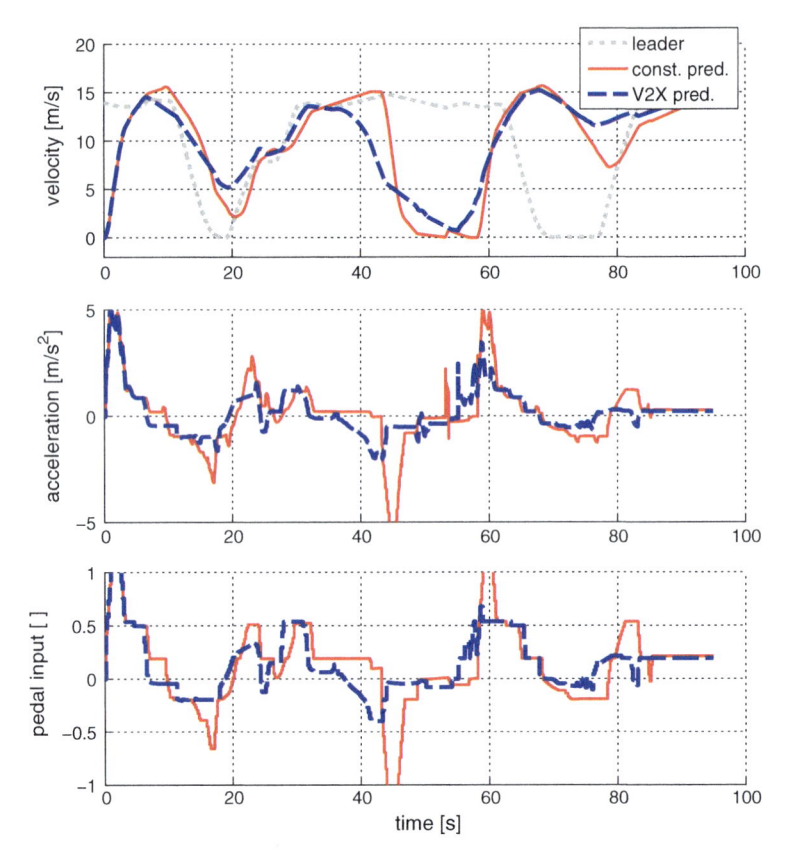

Fig. 6.10 Comparison of state trajectories for P-ACC with constant and V2X-based prediction of leading vehicles

6.4.3 Advantages of the Proposed Framework Versus Single Tools

TraSim only

If only the microscopic TraSim is utilized, it is possible to test the prototype driving function by setting the position or velocity profile of the ego vehicle. A disadvantage is the simplified vehicle model which is used in this simulations. Thus, an additional vehicle model in MATLAB may be required. Analysis of the potential impacts on the powertrain, such as consumption, emissions, or driving comfort is not directly possible. This would be a drawback for the presented P-ACC case study.

Additionally, complex dynamic interactions, like undesired oscillations, which may be caused by the ADAS in the vehicle cannot be tested. If the considered ADAS is expected to operate in highly transient and dynamic vehicle motion (e.g., automated emergency braking—AEB), the high-fidelity vehicle model is crucial. An advantage is the potential to investigate the impact on the overall traffic flow and the possibility to use advanced actuated traffic control methods.

VehSim only

An opposing remark can be made, if only the VehSim is considered. In this case, it would be necessary to set up the trajectories and maneuvers of all other traffic participants including traffic signal timing. The analysis of the powertrain and ego vehicle-dependent performance criteria would be possible. If a large variation of test cases in varying traffic conditions is necessary, then for all scenarios, the trajectories of the vehicles have either to be predefined or controllers developed for them.

Proposed framework

As mentioned the proposed framework aims to combine the advantages of both tools and so improve the development and testing process. It is possible to test a large number of scenarios rather easily, e.g., by varying traffic density or starting coordinates and time of the ego vehicle or traffic parameter distributions and vehicle classes in PTV VISSIM. The introduced EnSim layer provides an easy to use direct interface for virtual sensors and prototype ADAS functions and allows to develop the controller in MATLAB-Simulink.

6.5 Summary

A framework for virtual development and evaluation of ADAS has been presented. It combines a high detail vehicle simulation tool with a traffic simulation environment and a control development framework. A co-simulation framework is set up which combines both tools and handles the information exchange on the different levels. Therefore, an interfacing layer, EnSim, is introduced which additionally serve as interface for virtual sensors. The main focus is on a functional and high-level description to make studying of interaction between ADAS and other traffic participants possible. On an environment level all information is shared directly, whereas on a ADAS function level, virtual sensors are employed to extract the limited information available on a function level, similar to a real vehicle.

As a case study for the ADAS, a P-ACC is presented where the potential of the approach could be evaluated within the proposed framework. Although the case study is based on some simplifications and assumptions, it shows the potential of both the proposed ADAS utilizing information from V2X and I2V and the evaluation within the framework.

The current framework presents a work in progress of the evaluation framework. Additions, such as taking into account the traffic control based on available V2X and I2V information or more sophisticated sensor models, are currently under development. Another important step is the move to a full vehicle in the loop setup with a real test vehicle on a powertrain test bed which is part of the ongoing research project.

Acknowledgements This work has been funded by FFG under the project *Traffic Assistant Simulation and Traffic Environment—TASTE*, project number 849897. The authors further want to thank and acknowledge the whole TASTE project team for the assistance, fruitful discussions, and suggestions during the course of this work.

References

1. Barceló, J., et al.: Fundamentals of Traffic Simulation, vol. 145. Springer, New York (2010)
2. Bemporad, A., Morari, M.: Control of systems integrating logic, dynamics, and constraints. Automatica **35**(3), 407–427 (1999)
3. Bengler, K., Dietmayer, K., Farber, B., Maurer, M., Stiller, C., Winner, H.: Three decades of driver assistance systems: review and future perspectives. IEEE Intell. Transp. Syst. Mag. **6**(4), 6–22 (2014). https://doi.org/10.1109/MITS.2014.2336271
4. Bham, G.H., Benekohal, R.F.: A high fidelity traffic simulation model based on cellular automata and car-following concepts. Transp. Res. Part C Emerg. Technol. **12**(1), 1 – 32 (2004). https://doi.org/10.1016/j.trc.2002.05.001. http://www.sciencedirect.com/science/article/pii/S0968090X03000573
5. Brackstone, M., McDonald, M.: Car-following: a historical review. Transp. Res. Part F Traffic Psychol. Behav. **2**(4), 181–196 (1999). https://doi.org/10.1016/S1369-8478(00)00005-X
6. CarMaker, I.: Users guide version 4.5. 2. IPG Automotive, Karlsruhe, Germany (2014)
7. Fellendorf, M., Vortisch, P.: Microscopic Traffic Flow Simulator VISSIM, pp. 63–93. Springer, New York (2010). https://doi.org/10.1007/978-1-4419-6142-6_2
8. Festag, A.: Cooperative intelligent transport systems standards in europe. IEEE Commun. Mag. **52**(12), 166–172 (2014). https://doi.org/10.1109/MCOM.2014.6979970
9. Friedrich, B.: The effect of autonomous vehicles on traffic. In: Autonomous Driving, pp. 317–334. Springer, Berlin (2016)
10. Fritzsche, H.: A model for traffic simulation. Traffic Eng. Control **35**(5), 317–321 (1994)
11. Griggs, W.M., Ordez-Hurtado, R.H., Crisostomi, E., Husler, F., Massow, K., Shorten, R.N.: A large-scale sumo-based emulation platform. IEEE Trans. Intell. Transp. Syst. **16**(6), 3050–3059 (2015). https://doi.org/10.1109/TITS.2015.2426056
12. Helly, W.: Simulation of bottlenecks in single lane traffic flow. In: Proceedings of the Symposium on Theory of Traffic Flow, pp. 207–238. General Motors Research Laboratories, Elsevier (1959)
13. Kaths, J., Krause, S.: Integrated simulation of microscopic traffic flow and vehicle dynamics. In: IPG Apply & Innovate 2016. Karlsruhe (2016)
14. Krajzewicz, D., Erdmann, J., Behrisch, M., Bieker, L.: Recent development and applications of SUMO—Simulation of Urban MObility. Int. J. Adv. Syst. Meas. **5**(3&4), 128–138 (2012)
15. Krauß, S.: Microscopic modeling of traffic flow: Investigation of collision free vehicle dynamics. Ph.D. thesis, Universitat zu Köln (1998)
16. Lang, D., Stanger, T., Schmied, R., del Re, L.: Predictive cooperative adaptive cruise control: fuel consumption benefits and implementability. In: Optimization and Optimal Control in Automotive Systems, pp. 163–178. Springer, Cham (2014)
17. Li, L., Huang, W.L., Liu, Y., Zheng, N.N., Wang, F.Y.: Intelligence testing for autonomous vehicles: a new approach. IEEE Trans. Intell. Veh. **1**(2), 158–166 (2016). https://doi.org/10.1109/TIV.2016.2608003
18. Li, N., Oyler, D., Zhang, M., Yildiz, Y., Girard, A., Kolmanovsky, I.: Hierarchical reasoning game theory based approach for evaluation and testing of autonomous vehicle control systems.

In: 2016 IEEE 55th Conference on Decision and Control (CDC), pp. 727–733 (2016). https://doi.org/10.1109/CDC.2016.7798354
19. Löfberg, J.: YALMIP: a toolbox for modeling and optimization in MATLAB. In: Proceedings of the CACSD Conference. Taipei, Taiwan (2004). http://users.isy.liu.se/johanl/yalmip
20. Maurer, M., Lenz, B., Gerdes, J.C., Winner, H.: Autonomes Fahren: technische, rechtliche und gesellschaftliche Aspekte. Springer Vieweg (2015)
21. Moser, D., Waschl, H., Schmied, R., Efendic, H., del Re, L.: Short term prediction of a vehicle's velocity trajectory using ITS. SAE Int. J. Passeng. Cars **8**(2), 364–370 (2015)
22. N.N.: VISSIM Driving Simulator Interface - User Manual for Vissim 8. PTV Group (2017)
23. Schmied, R., Moser, D., Waschl, H., del Re, L.: Scenario model predictive control for robust adaptive cruise control in multi-vehicle traffic situations. In: 2016 IEEE Intelligent Vehicles Symposium (IV), pp. 802–807 (2016). https://doi.org/10.1109/IVS.2016.7535479
24. Stellet, J.E., Zofka, M.R., Schumacher, J., Schamm, T., Niewels, F., Zllner, J.M.: Testing of advanced driver assistance towards automated driving: a survey and taxonomy on existing approaches and open questions. In: 2015 IEEE 18th International Conference on Intelligent Transportation Systems, pp. 1455–1462 (2015). https://doi.org/10.1109/ITSC.2015.236
25. Wheeler, T.A., Kochenderfer, M.J.: Factor graph scene distributions for automotive safety analysis. In: 2016 IEEE 19th International Conference on Intelligent Transportation Systems (ITSC), pp. 1035–1040 (2016). https://doi.org/10.1109/ITSC.2016.7795683
26. Wiedemann, R.: Simulation des verkehrsflusses. Schriftenreihe des Instituts für Verkehrswesen, Universität Karlsruhe (8) (1974)
27. Zhao, D., Lam, H., Peng, H., Bao, S., LeBlanc, D.J., Nobukawa, K., Pan, C.S.: Accelerated evaluation of automated vehicles safety in lane-change scenarios based on importance sampling techniques. IEEE Trans. Intell. Transp. Syst. **18**(3), 595–607 (2017). https://doi.org/10.1109/TITS.2016.2582208
28. Zhou, J., Schmied, R., Sandalek, A., Kokal, H., del Re, L.: A framework for virtual testing of ADAS. SAE Int. J. Passeng. Cars Electron. Electr. Syst. **9**(2016-01-0049), 66–73 (2016)
29. Zofka, M.R., Klemm, S., Kuhnt, F., Schamm, T., Zöllner, J.M.: Testing and validating high level components for automated driving: simulation framework for traffic scenarios. In: Intelligent Vehicles Symposium (IV), 2016 IEEE, pp. 144–150. IEEE (2016)

Chapter 7
A Vehicle-in-the-Loop Emulation Platform for Demonstrating Intelligent Transportation Systems

Wynita Griggs, Rodrigo Ordóñez-Hurtado, Giovanni Russo and Robert Shorten

Abstract In an emerging world of large-scale, interconnected, intelligent transportation systems, demonstrating and validating novel ideas and technologies can be a challenging one. Traditionally, one is presented with a choice to make, between performing demonstrations with a few proof-of-concept "outfitted" vehicles, or experimenting with large-scale computer simulation models. In this chapter, we revisit a recent vehicle-in-the-loop (VIL) emulation platform that was developed with the goal in mind of taking steps towards countering the above validation dilemma. Roughly speaking, it was shown that a real, outfitted test vehicle, equipped with novel intelligent transportation technologies, could be "embedded" in a large-scale traffic emulation being performed with the microscopic traffic simulation package SUMO, thus allowing the real vehicle and driver to interact with thousands of simulated cars on a common road map in real-time. In our present work, we now provide an overview of the latest updates to the VIL platform, which include some enhancements to increase the platform's versatility and improve its functionality.

W. Griggs (✉) · R. Ordóñez-Hurtado · R. Shorten
School of Electrical, Electronic and Communications Engineering,
University College Dublin, Belfield Dublin 4, Ireland
e-mail: wynita.griggs@ucd.ie

R. Shorten
e-mail: robert.shorten@ucd.ie

R. Ordóñez-Hurtado · G. Russo
IBM Research—Ireland, Optimisation and Control Group,
IBM Dublin Technology Campus, Building 3, Damastown Industrial Estate,
Mulhuddart Dublin 15, Ireland
e-mail: rodrigo.ordonez.hurtado@ibm.com

G. Russo
e-mail: grusso@ie.ibm.com

7.1 Introduction

Vehicle-in-the-loop (VIL) simulation is emerging as an increasingly popular tool by which to circumvent a common dilemma arising in regard to the demonstration and validation of novel intelligent transportation technologies. This dilemma concerns the fact that intelligent transportation systems (ITS) are often intended to be deployed in large urban areas or major cities, and thus gaining access to fleets of thousands of vehicles equipped with the prototype technologies and communications abilities necessary for demonstrating ITS is usually not practical nor easily achievable. On the one hand, simulators can be used to compensate and emulate large scale, but cannot accommodate for all of the unmodelled vehicle dynamics, and other complexities, uncertainties, technical issues, and driver attitudes and responses that might arise in the real world [1]; not to mention that, given the rapid development and deployment of ITS, the associated experience of being in a connected vehicle scenario for many drivers will be brand new. On the other hand, small, real-world test fleets of one to twenty vehicles can demonstrate proof-of-concept, but cannot accurately predict the outcomes of ITS applied in the context of much larger fleet sizes and city-wide scenarios [2].

In a recent paper [2], we explored a low-cost, relatively straightforward method of merging large-scale traffic simulation and the proof-of-concept capability provided by real-world vehicles. A VIL platform for embedding, in real time, a real vehicle into Simulation of Urban MObility (SUMO) was built. SUMO is an open-source microscopic road traffic simulation package primarily being developed at the Institute of Transportation Systems at the German Aerospace Centre (DLR) [3]. Utilising our VIL platform, we then demonstrated a number of experimental ITS applications that we had developed primarily at the National University of Ireland, Maynooth in Maynooth, Ireland. These applications were examples of large-scale feedback systems built upon infrastructure-to-vehicle (I2V) capabilities. A goal of our demonstrations was to illustrate how to provide drivers of real vehicles the opportunity to somewhat experience what it would feel like to be part of a large-scale connected vehicle scenario, where the rest of the traffic in the scenario was simulated by SUMO in order to avoid the necessity of large, real vehicle test fleets. In addition, our platform aims to: (i) permit feedback of real vehicle and driver responses that, in pure simulation alone, may be unmodelled, unpredicted or unexpected (e.g. delays in reaction time and imprecise vehicle control); (ii) be inexpensive to build and (iii) illustrate potential for performing safe field demonstrations (e.g. on empty roads or test tracks) while scenarios such as widespread traffic congestion or accidents are simulated.

In this present work, we now aim to provide an overview of the latest updates to our VIL platform; see Sect. 7.4. Our updates include enhancements that aim to increase the platform's versatility and improve its functionality. We accompany the descriptions of these updates with some discussion on the potential ITS applications that our VIL platform could now demonstrate and validate; in particular, a more in-depth example of a novel speed advisory system is provided in Sect. 7.5. First, however, we revisit the basic architecture of our VIL platform in Sect. 7.3 and explore some state of the art in regard to hardware-in-the-loop (HIL) and VIL simulation in general in Sect. 7.2. Our future objectives are discussed in Sect. 7.6.

7.2 State of the Art

HIL simulation provides a means by which to add the complexity of a real plant under control to a virtual testing platform. Circumstances that can warrant the use of HIL simulation include: tight development schedules; practicality; development program cost savings; test safety and the need to consider real humans and/or systems in the loop. HIL simulation has had a place in the automotive industry for quite some time, for instance, in the testing of automotive anti-lock braking systems [4], diagnostic software [5] and electric vehicle drive trains and controllers [6]. Recently, however, in addition to the established focus on testing new systems intended for integration into a single vehicle, in models of said vehicles, the concept of HIL simulation has been expanded upon such that it also encompasses the testing of entire vehicles, equipped with Advanced Driver Assistance Systems (ADAS) or ITS technologies (along with real drivers or autonomous driving capabilities), in potentially very large scale or many multiple different traffic scenario simulations. The term being used for this kind of testing is vehicle-in-the-loop (VIL) simulation; see, for example, [7].

7.2.1 VIL Simulation Platforms

A number of VIL simulation platforms for testing ADAS currently exist. Examples include those described in [8–12], as well as in [2] and its predecessor [13]. The work in [9] particularly focused on autonomous intervening assistance systems for collision avoidance or mitigation. A VIL test setup was described in order to build upon existing validation methods such as driving simulators, traffic flow simulations and test vehicles that collide with substitutes such as foam cubes; see, also, [14]. Similarly, with respect to accident mitigation and avoidance, it was observed in [8] that VIL testing had a place, not only in validating the technical functionality of intervening systems, but also in studying the behaviour of real drivers as they interacted with the new technology. VIL was proposed as a method to combine virtual, visual simulation with the kinesthetic, vestibular and auditory feedback of a real car driving on a real test track, and thus VIL was promoted as capable of offering a variety of new options for evaluating ADAS, and of providing a real driving experience combined with the safety and test replication abilities of driving simulators. See, also, [15].

The work of [16] presents a midway approach to ADAS prototyping and validation, laying somewhere between HIL and VIL, that the authors refer to as vehicle hardware-in-the-loop (VeHIL). Their particular system is called SERBER. As opposed to having a driver travelling on a real test track, as was the case in [8], real vehicles in the setup of [16] are physically locked on a chassis dynamometer, and thus, the tests are conducted indoors. In this setup, environmental parameters such as humidity, ambient light, temperature can be easily controlled. The chassis dynamometer is paired with multi-sensor road environment simulation software. TASS International offers a VeHIL laboratory; see [17].

In [18], the concept of subsystem development occurring in parallel across different parts of the globe (for example, in globally distributed company departments) was tackled via the introduction of a new validation concept called X-in-the-distance-loop. Typically, bringing such subsystems together physically in order to validate interactions between them, as well as the system's behaviour in total, requires effort in terms of time and costs in regard to transportation. Furthermore, additional problems concerning confidentiality may need to be taken into account. It was proposed in [18] to utilise Internet connectivity in order to perform validation experiments from distributed geographical locations. As such, capabilities for real-time data transfer capabilities had to be realised, and the required information from each side defined by the developers and testers together.

The objective in [19] was to implement a VIL platform, where intersection control policies for autonomous vehicles, formerly only tested in simulation, were tested with a real autonomous vehicle that interacted with multiple virtual vehicles, at a real intersection, in real time. In such experiments, having all real autonomous vehicles (as opposed to a single, real autonomous vehicle interacting with a number of virtual vehicles) would have proven expensive, especially in the event that any of the control policies were to fail and an accident ensued. At the same time, the experiment yielded results that differed to those obtained when using a fully simulated environment.

GrooveNet, a vehicle-to-vehicle (V2V) network simulator capable of achieving communication between simulated vehicles, real vehicles, and between real and simulated vehicles, was described in [20]. With this approach, it became feasible to deploy a small fleet of vehicles in the field (e.g. in the order of a dozen) to test protocols that in truth involved hundreds or thousands of vehicles, the rest of which were simulated. GrooveNet was designed to investigate V2V issues, in particular, with respect to wireless communication issues in mobility networks.

We conclude this section by drawing attention back to our own work described in [2]. This VIL platform was developed with low cost and construction simplicity in mind, in addition to the idea of offering human drivers in real vehicles, on real roads or test tracks, opportunities to test and validate new ITS technologies in large-scale, interconnected, simulated traffic scenarios. In what follows, we briefly review our platform's existing architecture. We will then discuss the latest updates to our work and provide example applications.

7.3 Platform Architecture

The first iteration of our VIL platform, described in [2], was constructed as a preliminary prototype. Its setup was simple in design and consisted of an open-source microscopic traffic simulation package, emulating potentially thousands of virtual vehicles, sitting on a workstation computer in a control room; together with some ITS applications, also deployed on the workstation computer, written in Python. A real vehicle was "embedded" into our traffic simulations, the real vehicle being represented in the emulations by an avatar; and data was transferred between the real vehicle and the workstation computer in the control room, over a cellular network via a smartphone carried with the real driver. Some simple TCP-based client/server programming sat on both the smartphone side and workstation computer side. Our open-source traffic simulation package of choice was SUMO [3]. This simulator is designed to handle large road networks and comes with a "remote control" interface, TraCI (short for Traffic Control Interface) [21] that allows one to adapt the simulation and to control singular vehicles on the fly. An illustration of the platform setup that was described in [2] is presented in Fig. 7.1.

Some experimental ITS applications, developed primarily at the National University of Ireland, Maynooth in Ireland, that we then proceeded to demonstrate using our VIL platform, in [2], consisted of the following: (a) an intelligent speed recommender system [22]; (b) emissions regulation, via context-aware hybrid vehicle engine mode control [23] and (c) local rerouting around an obstruction [24]. As mentioned, Python scripts containing algorithms that were unique to each specific ITS application were deployed on the workstation computer. These Python scripts additionally contained some transmission control protocol (TCP) server code for handling connections with, and data flow to/from, the smartphone, as well as some code that enabled our scripts as clients which communicated with SUMO, via TraCI, the traffic simulator itself thus performing as a server. (TraCI uses a TCP-based client/server architecture to provide access to SUMO.) These latter two aspects of our scripts (i.e. the TCP-based components) were universal with respect to all three of our ITS applications.

Let us review the setup of the platform with regard to two of these demonstrated ITS applications more closely.

7.3.1 Intelligent Speed Recommender

The goal of this application was to detect approaching traffic bottlenecks, e.g. roadwork zones or traffic jams, and provide drivers with recommended travelling speeds, ensuring that vehicles travelled at safe speeds and distances from the vehicles ahead of them as they approached, entered and left the bottleneck. The notion was that a traffic bottleneck could be emulated on demand in the virtual environment, while proof-of-concept of the system was being demonstrated in the real vehicle on an empty road. Validating the system with the VIL platform included obtaining an initial assessment

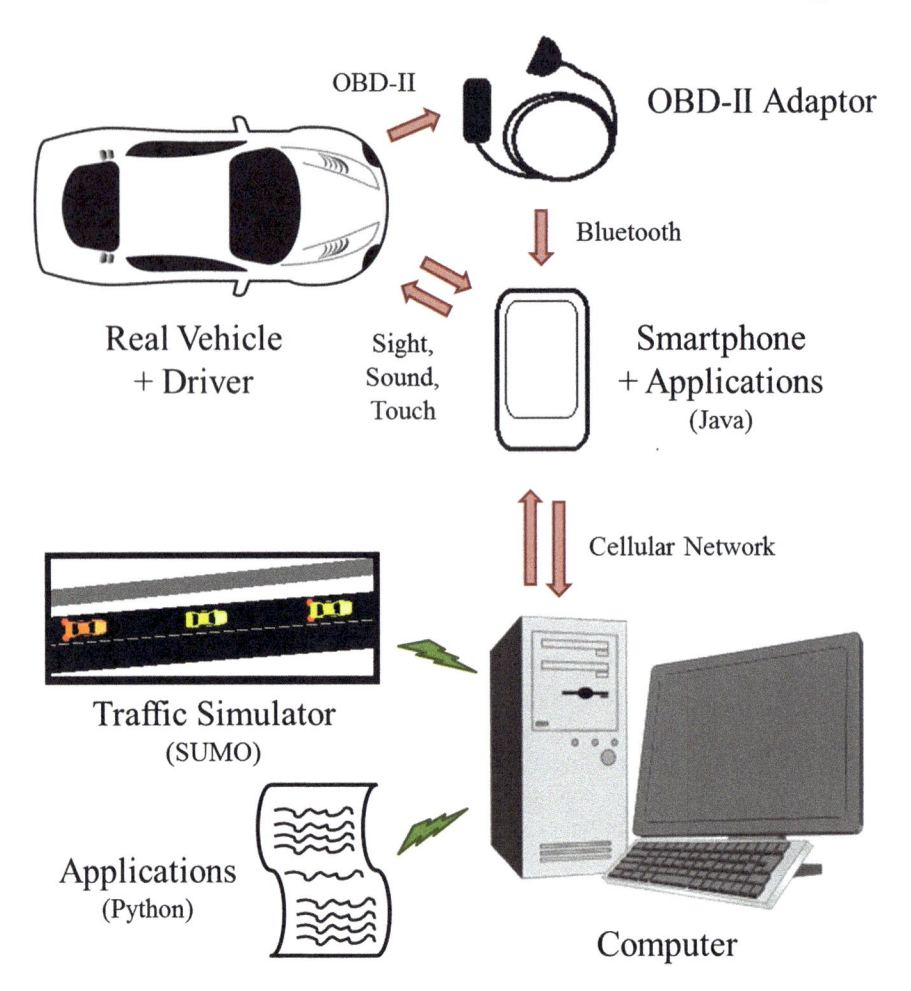

Fig. 7.1 Illustration of the VIL platform setup described in [2]. In [2], the essential notion was that ITS applications were deployed on the workstation computer also interfacing hundreds of vehicles in SUMO, while a corresponding application was deployed on the smartphone which served as the interface to the driver and the real vehicle. The simulated vehicles and the real vehicle, both provided inputs for the control algorithms and applications to be tested. (The real vehicle and computer sub-images contained within this illustration were downloaded from https://openclipart.org on 27 July 2016 and 22 September 2017, respectively.)

of a real driver's attitudes and responses to the speed recommendations, as well as demonstrating how feedback from the real driver and vehicle could be incorporated into the evolving traffic situation.

The application algorithm consisted of two key stages. First, the traffic scenario in which the *host vehicle* (i.e. the vehicle of interest, receiving the recommendations) was travelling in was determined. A full list of potential traffic scenarios consisted of: free traffic; approaching congestion; congested traffic; passing bottleneck and

7 A Vehicle-in-the-Loop Emulation Platform for Demonstrating ... 139

Fig. 7.2 Field test vehicles

leaving congestion. An estimate of the vehicle density surrounding and information concerning the speeds of the host vehicle and the *next vehicle* (i.e. a point of interest along the future trajectory of the host vehicle) were utilised by a rule-based inference engine in order to determine the traffic scenario. Then, using this (and some additional) information, a recommended travelling speed for obtaining a safer distance to the next vehicle was calculated.

In our setup, in [2], the host vehicle was our real vehicle. For this vehicle, we utilised a 2008 Toyota Prius 1.5 5DR Hybrid Synergy Drive, pictured on the left in Fig. 7.2; however, any vehicle with an accessible interface or gateway to its onboard computer, such as an OBD-II diagnostic connector, would have been suitable to use. In the vehicle, we mounted a Samsung Galaxy S III mini (model no. GT-I8190N) running the Android Jelly Bean operating system (version 4.1.2). The purpose of the smartphone was to relay, over a cellular network, periodic updates from the vehicle's onboard computer in relation to its current speed, to the workstation computer in the control room running SUMO, and to receive the recommendation messages from the workstation computer and display them on a user interface for the driver. We utilised the mobile data services of a commercial mobile phone operator for the relay of data, where these services were provided using a 3G UMTS 900/2100 network.

The hardware device that we used to form the connection between the smartphone and the Prius' onboard computer (via its OBD-II diagnostic connector) was

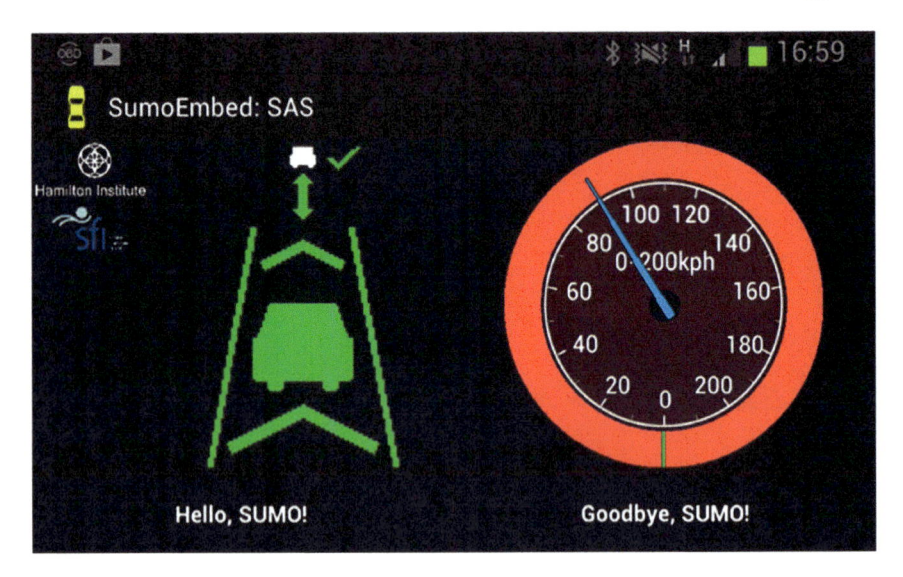

Fig. 7.3 Plug-in user interface for the speed recommender system demonstrated in [2]. Adapted from [2], © 2015 IEEE, used with permission

a Kiwi Bluetooth OBD-II Adaptor by PLX Devices. This device was plugged into the vehicle's OBD-II diagnostic connector and communicated the data relating to the vehicle's current speed to the smartphone via Bluetooth. A variety of existing smartphone applications were compatible for use with the Kiwi Bluetooth at the time of our experiment. We decided upon Torque Pro[1] given that an Android Interface Definition Language (AIDL) application programming interface (API) was included with it for the development of third-party plug-ins. We utilised this feature to design a new plug-in for Torque Pro for our speed advisory application. The functionalities of this plug-in included: buttons for initiating and terminating communication with the workstation computer, and the associated Java socket programming for maintaining this communication and handling all of the data transfer; the capability to bind to the Torque Pro service running in the background, and thus extract information in relation to the vehicle's current speed from the Prius' onboard computer; and a unique user interface that showed to the driver the vehicle's current speed, as well as the traffic scenario that the algorithm running on the workstation computer had determined that the Prius was currently travelling in, and the speed recommendation that the algorithm consequently issued to the driver. This user interface is shown in Fig. 7.3.

We also created a remote procedure call (RPC) framework based on TCP sockets, which allowed us to directly call MATLAB functions from our Python script that was deployed on the workstation computer. We did this because, for this particular

---

[1] *Torque Pro* by Ian Hawkins. Available from Google Play: https://play.google.com/store/apps/details?id=org.prowl.torque. Last accessed on 28 September, 2017.

demonstration, our ITS algorithm was initially developed using the MATLAB Fuzzy Logic Toolbox (Version 2.2.14, R2011b). Thus, by utilising the RPC framework, we were not forced to port the algorithm, developed in MATLAB, over to Python.

For our demonstration then, the Toyota Prius was driven around a circuit defined on the north campus of the National University of Ireland, Maynooth. The demonstration was performed at a time of day when the rest of the real traffic on the road was minimal (i.e. early in the morning). The road map generated for use in SUMO and the real road circuit that the Prius travelled upon were topographically the same. Twenty-two virtual vehicles with different characteristics (i.e. maximum speeds, sizes, acceleration and deceleration capabilities) were emulated in SUMO, in real time, during the Prius' test drive, along with a variety of local speed limits (lower than the real speed limits on the road), in order to create bottlenecks of traffic that involved both the simulated vehicles and the Prius, which was represented in the emulation by an avatar. At the beginning of the test, both the Prius and its avatar in the simulation departed from the same positional starting point on the circuit. Updates to the Prius' position in the SUMO map were then made based (only) on its real-time travelling speed as obtained through its OBD-II diagnostic connector. Given that its real GPS coordinates, for example, were not used in making updates to its position in the SUMO map, map-matching with respect to representing the Prius' actual position (on the real road) accurately in the SUMO map was poor. Thus, it was noted that the relay of real-time positional data would be critical for much future work, and that better map-matching was an important element yet to be included in the platform. Other results and observations from the demonstration, e.g. in relation to the driver's comfort at following the speed recommendations, were reported in [2].

7.3.2 Emissions Regulation

In this ITS application, the different engine modes available to hybrid vehicles were utilised to regulate emissions levels, in a fair manner for all participating vehicles, over geographical zones where maximum levels of tolerated pollution were applied. Specifically, the application sought to orchestrate the way in which each vehicle, from a large fleet of hybrids, would uniquely utilise its internal combustion versus electric engine modes to regulate pollution, through simple communication signals from a central infrastructure. A number of algorithms were provided in [23] as a means of achieving this orchestration. In [2], the simplest of these algorithms was demonstrated; namely, a simple integral controller was applied in a stochastic framework.

Specifically, a threshold was set in regard to what the permissable level of global CO emissions to be tolerated during our traffic scenario emulation would be; and then, feedback regulation was employed by a central infrastructure, which sent periodic signals to each of the hybrid vehicles participating in the scenario, such that the global CO emissions in the scenario would be regulated around the threshold in a manner that was fair to all participating hybrids. The signals that were sent by the central infrastructure pertained to a probability value, as follows. In our scenario, internal combustion engine use resulted in a vehicle emitting CO, while driving in fully electric mode did not. Engine mode instruction for each unique vehicle, at regular intervals in time, was dictated by a periodic "coin flip", the result of this "coin flip" (i.e. random number generation) being compared to the aforementioned probability value. This probability value reflected what the global CO emissions for the scenario were at the current moment. If current CO levels for the scenario were zero, then the probability of any participating hybrid vehicle being allowed to wholly utilise its internal combustion engine, for example, was one. This probability degraded towards zero as global CO emissions in the scenario rose. The results of the demonstration are provided in [2]. We describe, next, further specifics of the demonstration setup in regard to our VIL platform.

It was assumed that we had access to all necessary environmental information for our demonstration, such as the exact amounts of CO being emitted by each vehicle (both virtual and real cars; this information was derived using the formulae cited in [23]); and we set an artificial regional pollution threshold. For our real vehicle that was embedded in the emulation to provide the proof-of-concept aspect of the demonstration, we again elected to use the 2008 Toyota Prius 1.5 5DR Hybrid Synergy Drive, as pictured on the left in Fig. 7.2. Again, we utilised the Samsung Galaxy S III mini (model no. GT-I8190N) to relay, over the cellular network, periodic updates from the vehicle's onboard computer, via the Kiwi Bluetooth OBD-II Adaptor, in relation to its current speed (this speed information was utilised on the workstation computer to compute the Prius' CO emissions via the formulae cited in [23]), and to receive messages from the workstation computer in relation to engine mode orchestration for the next time step (i.e. to receive the current global probability value arising from the current global CO level).

The plug-in for Torque Pro that we developed for the application again consisted of buttons for initiating and terminating communication with the workstation computer, and the associated Java socket programming for maintaining this communication and handling all of the data transfer, and again had the capability to bind to the Torque Pro service running in the background, to extract information in relation to the vehicle's current speed from the Prius' onboard computer. The user interface again showed the driver of the Prius its current speed; and this time also displayed a message in relation to what engine mode the driver of the Prius should currently be employing, after the plug-in performed one of its "coin flips" and compared the random number obtained to the corresponding most recent probability value received from the workstation computer. The user interface additionally consisted of two buttons that allowed the driver to engage or disengage automatic engine mode control. When automatic engine mode control was engaged, signals were forwarded from the smartphone, via Bluetooth, to a mechanical "finger" device that was mounted inside the Prius. The "finger" physically interacted with the Prius by pushing and releasing a drive mode switch in the vehicle. This mechanical device is shown in Fig. 7.4. The user interface of the plug-in was shown in [2].

Given that the control algorithm itself had been initially developed in MATLAB, we again made use of our RPC framework to directly call MATLAB functions from

Fig. 7.4 Mechanical device

our Python script that was deployed on the workstation computer. As such, we were again not forced to port the algorithm, developed in MATLAB, over to Python. For convenience, we also utilised, in the demonstration, the same road map and SUMO configurations (with respect to the number and type of virtual vehicles in the emulation) that we did for the speed recommender application. Again, updates to the Prius' position in the SUMO map were made based (only) on its real-time travelling speed.

In our demonstration, we applied a constant threshold, in regard to permissable global CO emissions, over the entire map. However, we recognised that obtaining positional information from the real car (e.g. GPS coordinates) would be necessary if a constant threshold was not applied over the whole map, but rather to subregions. In such cases, positional data would be needed from the vehicle to determine accurately whether or not it was inside a region where a pollution threshold was enforced. Furthermore, we noted that incorporating event-driven logic into our Python script and plug-in, to activate the application when the real vehicle entered a geographical zone where pollution monitoring was ongoing, would be ideal. Finally, we noted that, in reality, the exact quantity of CO being emitted by each vehicle would not be known to a central infrastructure. Furthermore, in more realistic scenarios, not only hybrid vehicles participating in the service would be driving around on the map. So too, driving around, would be non-participating hybrids; and also vehicles that were not hybrids and thus could not have their engine modes affected by the application. Therefore, knowledge of CO levels in a region would likely come from external measurements, for example, from a meteorological agency utilising sensors. To imitate such in our VIL platform, we noted that another ideal future enhancement to the platform would be an ability to incorporate real-time and/or real-world information in order to bring more realism or proof-of-concept capabilities to our demonstrations; for example, live news events, traffic jam reports and weather and pollution information from meteorology organisations could be taken into account in generating or validating test scenarios.

7.3.3 On the Existing Architecture

While we succeeded, in [2], in developing a preliminary prototype of our VIL platform, and demonstrated some ITS applications, and thus exhibited the concept of the VIL platform that we aimed to construct, it is clear from the discussion above that a number of ideal components or enhancements to the VIL platform remained yet to be added. In particular:

(i) Our original design permitted only a single, real vehicle to be embedded into the platform (this issue was not discussed above but will be explored in Sect. 7.5);
(ii) Access to real-world information was limited to what the real vehicle's on-board diagnostics (OBD-II) and/or what the smartphone, carried on board the real vehicle with the driver, could provide;
(iii) Our Python scripts were not modular, in that they did not exhibit a plug-and-play structure, in regard to what ITS applications were required by a user at any given point in time;
(iv) Our Python scripts were not event-driven, with respect to what ITS applications were in use, or active, in a traffic scenario at any given point in time and
(v) Our platform exhibited poor map-matching in regard to the incorporation of real-world positional information from the embedded hardware into the platform.

Thus, in this current work, we now aim to report on some improvements that we have made to our VIL platform since its inception in [2, 13]. These improvements are as follows:

(I) Multiple real vehicles can now be embedded through threading and a ticketing system;
(II) Real-world information is available from a range of different sources;
(III) We have moved towards code modularity and event-driven capabilities and
(IV) We have improved map-matching for those applications that are location-aware.

In the next section, we describe these improvements in more detail. Our aim is to maintain, as much as possible, a low cost in regard to platform construction, and ease of availability in terms of the components required, while also extending upon the range of ITS applications that the platform may be used to demonstrate and validate.

7.4 New Components and Enhancements

An account of the latest components added to the VIL emulation platform follows. As mentioned, our aim is to, through the addition of new features and enhancements, increase the platform's versatility and improve its functionality.

7.4.1 Scalability: Multiple Real Vehicle Embedding

The original design of our platform permitted only a single real vehicle to be embedded into a SUMO simulation. To add more real vehicles, we exploited the fact that, in the parent process of our Python scripts, TraCI provides access to SUMO via a single port; and multiple real vehicles can be represented in SUMO by different vehicle IDs. Thus, multiple real vehicles can be embedded in a single simulation.

Threading and the use of a ticketing system were implemented in the server subprocess of our Python application scripts (i.e. the subprocess that communicates with the smartphones). Threading permits the handling of multiple client calls at once. The procedure was as follows. A single port is reserved for communication with the smartphones in our server subprocess. When a new real vehicle joins the network and wishes to utilise the ITS application under test, it communicates its presence via its smartphone to the server. Once the server has accepted the connection, the real vehicle is provided with a unique ticket for identification purposes; parameters associated with the real vehicle (e.g. its speed and location) are initialised and, from there, communication with the real vehicle is passed to and dealt with in a newly created, personal, background thread for continued interaction between the server and client. That is, the real vehicle can now send continued data regarding its state, and the server can pass back information continually, such as alerts and recommendations.

Finally, we note that this modification to the platform permits not only multiple real vehicles to be added to a single simulation such that they may all drive around in it concurrently, but also enables single real vehicles that leave the scenario and need to re-enter it again later, at a future time step, to be successfully re-added when (or each time that) they do. We will explore a benefit of being able to embed multiple real vehicles in the use case presented in Sect. 7.5.

7.4.2 Information Exchange: Additional Sensors and Other Devices

In [2], our platform architecture utilised the OBD-II diagnostic connector of a real vehicle to access data from its engine control unit (ECU). The particular data accessed is typically application-driven and might consist of the vehicle's speed, the state of charge of its high voltage battery, etc. The type and amount of data available to ITS application testing need not be limited to what the OBD system can provide, however. Real vehicles can be equipped with additional sensors and devices. Furthermore, other forms of real-world and/or real-time data can be obtained (for example, from meteorological organisations) and incorporated into testing, all in order to enable the available information about the real vehicle, driver and the real vehicle and driver's surroundings, to be expanded upon. Some of these additional devices and/or the data that they provide can be incorporated into the VIL platform and used for validating a wider range of ITS applications in a more comprehensive or realistic manner. Thus, we now discuss some new sensors and other devices that we have utilised recently, in terms of extending the capabilities of our VIL platform.

Smartphones

Besides acting as devices through which to relay, to a workstation computer over a cellular network, information obtained from a real vehicle via its OBD-II diagnostic connector, and as devices through which to receive messages from a workstation computer and display them on a user interface for a driver, smartphones have the capacity to provide other uses as well. By themselves, for example, they carry their own sensor compliments, such as GPS and accelerometers, and can thus provide extended information in addition to what is available from the real vehicle's ECU. They can receive and process human input (e.g. voice, touch or text) through a user interface, as provided by a driver. Smartphones have processing power and can run a range of applications. Some applications may involve the preprocessing of data to be sent to a workstation computer and/or involve computations, algorithms and actions concerning the data received from a workstation computer. Other applications (or subprocesses of an application) may involve processing information received from other, additional devices onboard a real vehicle, or carried by the driver, such as the wearable technology that we describe next.

Wearable Technology

With the rise of the Internet of Things (IoT), wearable technology (i.e. clothing and accessories incorporating computer and advanced electronics, such as Microsoft Band [25], Apple Watch [26] and Android Wear [27]) has become readily available. Wearables can provide additional information regarding the state of a real driver; for example, Microsoft Band and Apple Watch contain heart rate sensors. For instance, recently, at University College Dublin, in Belfield, Ireland, in collaboration with IBM Research, Ireland, one of our team's research projects has consisted of embedding an electric bicycle into a SUMO simulation and artificially creating regions where transport-related pollution is high (e.g. a busy section of road with many buses and trucks) [28, 29]. The bike is fitted with a controller that increases the output from the electric motor in order to aid the pedalling bike rider in powering the bike when the rider enters one of these "tail-pipe" polluted areas. The aim is to keep the bike rider's breathing rate lower so that they do not inhale as much of the air pollution. To monitor the heart and ventilation rates of the bike rider, they were fitted with ventilation masks (i.e. a COSMED Spiropalm 6MWT, with a fully integrated pulse oximeter) during the application testing, which took place both on a real road and in a laboratory. A video demonstrating the proposed system can be found at [30].

Gas Sensors and Weather Stations

In the emissions regulation example described in Sect. 7.3, we briefly mentioned the notion of incorporating real-world weather or pollution information, as measured by meteorology organisations, for example, in order to create more realistic test case scenarios or proof-of-concept demonstrations. Gas sensors, e.g. [31], can be fitted to real vehicles embedded in simulations for validation purposes too. As an alternative example application, consider [32], wherein the source localisation problem for a natural gas leak in an urban setting was considered. As an extension to this work, to provide proof-of-concept and improve upon the theoretical models presented within it, some members of our research team, in collaboration with IBM Research, Ireland, have since performed experiments using real CO_2 sensors [31] positioned in a two-dimensional space (as well as fitted on parked cars) and demonstrated gas leaks coming from different point sources. Experimentation with the sensors was possible due to the ease of incorporating the compact modules as add-on components of microprocessor-based equipment via serial communication, and also due to their technical features such as low power consumption, wide measurement range, fast response rate and high accuracy. The solution approach to the source location problem then involved using the information obtained from the CO_2 sensors. The measurements from the sensors were combined with wind information from weather stations and processed with a source localisation algorithm based on advection–diffusion equations [32]. It is worth noting that an off-the-shelf professional weather station, such as the one used in our investigations [33], not only provides wind information from a wind metre, but also incorporates a temperature sensor, humidity sensor, barometric pressure sensor, light sensor and a rain gauge, which allows it to be used in applications such as weather estimation and weather forecasting.

RFID

Radio frequency identification (RFID) is a generic term for technologies that use radio waves to automatically identify entities [34]. Several methods of identification exist, but the most common is to store a serial number that identifies an entity on a microchip that is attached to an antenna. The chip and antenna together are called an

RFID transponder or tag. The antenna enables the chip to transmit the serial number to a reader. The reader converts the radio waves reflected back from the RFID tag into digital information that can then be passed on to a computer to make use of. A passive RFID tag draws power from the field created by the reader and uses it to power the microchip's circuits. The chip modulates the waves sent back to the reader by the tag, and the reader converts these new waves into digital data. In this way, passive tags do not require a local power source. In addition, unlike barcodes, RFID tags do not need to be within the line of sight of the reader.

In some recent experiments, we utilised RFID in combination with our VIL platform to study problems relating to the localisation of missing entities by considering parked vehicles as service delivery platforms [35, 36]. Specifically, the notion was to utilise vehicles that are parked for extended periods of time in dense, urban areas to detect and track moving, missing objects using RFID technology. Such entities might consist of a missing patient with dementia, a lost pet or a stolen vehicle, e.g. a bicycle. The entities carry with them a passive RFID tag, while the readers are located with the parked vehicles. The advantages of using a network of parked vehicles to locate a missing entity include, first, the sheer number of vehicles that are owned by people, and the fact that, for 96% of the time on average, these vehicles are parked [37]. In other words, the network is large. The network also does not require dedicated maintenance, and technology upgrades are easy [38]. Additionally, energy infrastructure and planning permissions are not required to establish the network. The use of the VIL platform permits us to experiment with a range of algorithms; e.g. algorithms concerning the density of vehicles actively searching for an RFID tag at any point in time versus the time taken until localisation of the missing entity occur. One such algorithm was provided in [36]. Realistic parking data from urban environments can be utilised and represented in an emulated scenario, while proof-of-concept of the system in real life and in real time is being demonstrated. Preliminary results were provided in [36]. Further investigations are ongoing.

Another, related, recent use of RFID in research concerned the implementation of a low-cost, low-power, easily integrated cyclist collision prevention system for in-car deployment. The result of another collaboration between University College Dublin and IBM Research, Ireland, this system sought to inform car drivers of nearby cyclists, based on cyclists being equipped with passive RFID tags and cars containing readers. Further details can be found in [39]. Finally, we mention [40], wherein geo-fences (i.e. virtual geographic boundaries) were employed to specify areas of low pollution around cyclists. Similar to the emissions regulation application as discussed in Sect. 7.3, the emissions level inside a geo-fence was controlled via a coin tossing algorithm that determined the engine mode to be next employed by a participating hybrid vehicle. The algorithm was triggered when a vehicle detected a cyclist with RFID. The system was demonstrated with our VIL platform.

High-Precision GPS Information

We have utilised the high-precision, DGPS/AGPS-enabled, GPS travel recorder, Qstarz BT-Q1000XT [41], in our investigations to enhance the position estimation capabilities of real vehicles; and with the smartphone application Torque Pro, which permits the use of external (Bluetooth) GPS devices, we have been able to pass such information to the VIL emulation platform. This makes the localisation of avatars in SUMO a more reliable process, in terms of representing real-life events (i.e. the position of the embedded cars, in this case) with a higher level of accuracy.

7.4.3 Map-Matching: Speed and Position Corrections

The incorporation of real-world information from the embedded hardware into the VIL emulation platform should be performed as accurately as possible, so that the real-world information is represented with high fidelity in the simulated experiments. Two important variables obtainable from an embedded vehicle are its position and speed. Thus, we pay special attention to those next.

Concerning information pertaining to a real vehicle's speed, it is been reported that the speed measurements obtained from an ECU do not generally match the speed on the car dashboard display, and this has two main causes: (i) a typical (intentional) over-read in most standard speedometers and (ii) a different wheel diameter from the factory specification [42]. GPS speed measurements are also available, for example, from a smartphone, and optionally via Torque Pro, as an alternative.

In the case of position information, the main data source is GPS localisation. However, GPS measurements are highly sensitive to environmental and/or contextual factors, and they can be strongly degraded by atmospheric effects (e.g. ionospheric delay), multipath effects (e.g. from buildings), outdated ephemeris data and the GPS receiver quality (e.g. inefficient algorithms/circuitry). Accordingly, we have proposed different methods for mitigating the position errors from traditional GPS measurements. These methods include the use of:

1. A high-precision, DGPS/AGPS-enabled GPS receiver, e.g. [41], which delivers more accurate GPS measurements than the average smartphone's GPS unit;
2. Kalman filtering, which improves the position estimation accuracy by fusing GPS information (e.g. latitude, longitude, precision) with sensor data (e.g. speed from OBD-II, bearing from smart device gyroscope) and aggregated data (e.g. bearing estimation from previous GPS locations, speed from GPS);
3. Passive RFID tags, which are (intentionally) placed at dedicated roadside locations and can transmit their position information once they are detected by the car;
4. Map-matching, with respect to the road network used in the simulated environment, using the method https://traci.vehicle.moveToXY[2] and/or
5. Parked vehicles, used as anchor nodes for cooperative positioning (given the availability of V2V systems) [43].

---

[2]Under typical conditions, SUMO only allows vehicles to be on roads, and thus if an attempt is made to move a vehicle to an arbitrary location, then SUMO will place it on the nearest road to that location, taking into account certain filters. One of these filters uses bearing information so that the vehicle is map-matched to the nearest road with an orientation similar to the bearing provided by the user.

Of the above methods, (1), (2) and (3) have been successfully implemented and tested, while (4) is currently being implemented, and (5) is planned for a future design.

7.4.4 Transmission Frequency and Code Modularity: Application Logic

In the first iteration of our platform design, once a connection was formed, information between the real vehicle and the workstation computer was transmitted once per second. This steady, periodic rate was maintained throughout the duration of a simulation for simplicity. Changing the information transmission frequency, e.g. to different, event-inspired transmission frequencies, is something that makes sense for a number of applications, however. For instance, in the "detect and track a moving, missing object using RFID technology" application described in Sect. 7.4.2, one transmission frequency might be utilised while the detection stage of the application is active, while a different transmission frequency could be preferential once the object is located and the tracking function kicks in. Investigations regarding this are currently ongoing. As part of the ENABLE-S3 project [44] (an ongoing project funded by the ECSEL Joint Undertaking, which in turn receives support from the European Union's HORIZON 2020 research and innovation programme, as well as from a number of European countries), we are additionally continuing the development of the VIL platform such that ITS applications under test may be inserted in a "plug-and-play" capacity and activated, deactivated or modified in behaviour on demand.

7.5 A Final Illustrative Use Case

In this section, we conclude the discussion on our VIL platform improvements by returning to the notion of embedding multiple real vehicles into a single emulation. These vehicles could be embedded either concurrently or consecutively, the latter instance also being applicable to where there might be only a single real vehicle, but it is to exit and enter a scenario multiple times. We will now provide a use case example of a recently proposed speed advisory system (SAS) [45, 46] in which embedding multiple real vehicles (as opposed to a single one) is a necessary requirement for adequately validating the system when utilising the VIL platform. The reasoning behind this is presented below, in Sect. 7.5.1.

The objective of the SAS is to implement a consensus algorithm to guide a set of vehicles towards a common driving speed [46]. An innovation of the SAS is that consensus is achieved over a multi-layer network, where parallel network topologies of connected vehicles are superimposed. The reason for the use of these parallel networks is that, in this way, state obfuscation is possible, with the benefit that common driving speed is attained with no vehicle knowing the exact state of other vehicles participating in the service. The SAS can be demonstrated using our VIL platform, where we would have (for instance) the two real field test cars shown in Fig. 7.2 participating in the service along with many simulated vehicles. We direct the reader to [46] for further details on the SAS algorithms, and focus our attention here on the characteristics of the VIL demonstration instead.

In our setup, we considered the road network of the University College Dublin campus in Belfield, Ireland. A map of the campus was imported into SUMO from OpenStreetMap. A maximum allowed link speed of 30 km/h was set on the campus roads to reflect real-world speed limits, and a maximum allowed link speed of 50km/h was set on roads on the approach to the campus (i.e. on links on the approach to the campus entrance gates). The road network is shown in [46, Fig. 4]. In regard to the simulated traffic in our demonstration, we emulated a morning rush, with passenger vehicles randomly being allocated to enter one of three possible entrance gates, and then making a random decision to head to one of three possible car parks (see [46, Fig. 4]). Vehicles entered the network at a rate of one vehicle per twenty seconds, for twenty minutes, at maximum permitted departure speed. Other attributes that we assigned to the simulated vehicles can be found in [46].

In a certain test case scenario of interest to our current discussion, no vehicle (simulated or real) was instructed to be a leader, and thus speeds were expected to converge towards the average. The time step size used in the simulation was 0.1s; however, information was only exchanged every 1s between the workstation computer and the smartphones carried in the real vehicles. A VIL emulation was performed to test the convergence of the SAS.

7.5.1 Discussion

In the case where only a single real vehicle is embedded in the VIL platform while the SAS described above is under test, a potential arising issue is that the real vehicle can unintentionally become a *leader*. This is because a real driver's ability to follow speed advice is often less precise than a simulated car's ability. Thus, the real vehicle pulls the convergence of the algorithm in an up or down manner depending on its own speed value. Having multiple real vehicles embedded, which is now possible with our VIL platform, paves the way to negating this issue, supposing that all real drivers are attempting to follow the speed advice.

In addition, as mentioned prior, our platform now also permits for vehicles that leave a scenario and return to it later, to be "re-embedded" when (and each time) that they do. Not only is this an important feature for fault tolerance, like intermittent connection loss between a real vehicle and the workstation computer, but it also permits for us to accommodate situations, for example, where an application (like the above SAS) is active in a particular geographical zone, and real drivers are entering and exiting that zone multiple times throughout a demonstration. Each time that they do enter the zone, they can be reincorporated into the scenario as participants.

Finally, the methods for mitigating positional errors obtained from traditional GPS measurements that were described in Sect. 7.3 give us an improved capacity to map match and thus track the positions of the real vehicles on the road network in SUMO.

7.6 Conclusions and Future Work

The addition of new features and enhancements to our VIL emulation platform increases its versatility and improves its functionality in regard to ITS application demonstration and validating. Future features and enhancements that we would like to add include the capacity for our platform to be able to incorporate and handle V2V communications. To achieve this, the integration of our platform with a network simulator, and an ability of the platform to process information from real onboard units and roadside units (i.e. communication devices), will be necessary.

Acknowledgements This work was partially supported by Science Foundation Ireland grant 11/PI/1177.

References

1. Tielert, T., Killat, M., Hartenstein, H., Luz, R., Hausberger, S., Benz, T.: The impact of traffic-light-to-vehicle communication on fuel consumption and emissions. In: Internet of Things (IoT) Tokyo, Japan (2010)
2. Griggs, W.M., Ordóñez-Hurtado, R.H., Crisostomi, E., Häusler, F., Massow, K., Shorten, R.N.: A large-scale SUMO-based emulation platform. IEEE Trans. Intell. Transport Syst. 16(6), 3050–3059 (2015)
3. Krajzewicz, D., Erdmann, J., Behrisch, M., Bieker, L.: Recent development and applications of SUMO – Simulation of Urban MObility. Int. J. Adv. Syst. Meas. 5(3,4):128–138 (2012)
4. Hwang, T., Roh, J., Park, K., Hwang, J., Lee, K.H., Lee, K., Lee, S.-J., Kim, Y.-J.: Development of HILS systems for active brake control systems. In: SICE-ICASE International Joint Conference, pp. 4404–4408. Busan, Korea (2006)
5. King, P.J., Copp, D.G.: Hardware in the loop for automotive vehicle control systems development. In: UKACC Control Mini Sumposia, pp. 75–78. Bath, UK (2004)
6. Poon, J.J., Kinsy, M.A., Pallo, N.A., Devadas, S., Celanovic, I.L.: Hardware-in-the-loop testing for electric vehicle drive applications. In: 27th Annual IEEE Applied Power Electronics Conference and Exposition (APEC), pp. 2576–2582. Orlando, Florida, USA (2012)
7. Hammerschmidt, C.: Vehicle-in-the-loop speeds automotive design cycles. In: News. eeNews Europe Automotive (2015). http://www.eenewsautomotive.com/news/vehicle-loop-speeds-automotive-design-cycles Cited 11 Sep 2017
8. Berg, G., Nitsch, V., Färber, B.: Vehicle in the loop. In: Winner, H., Hakuli, S., Lotz, F., Singer, C. (eds.) Handbook of Driver Assistance Systems: Basic Information, Components and Systems for Active Safety and Comfort. Springer International Publishing Switzerland, Switzerland (2015)

9. Bock, T., Maurer, M., Färber, G.: Vehicle in the loop (VIL)—a new simulator set-up for testing Advanced Driving Assistance Systems. In: Driving Simulation Conference—North America (DSC–NA) Iowa City, Iowa, USA (2007)
10. IPG Automotive GmbH.: VIL systems. In: Products & Services. IPG Automotive (2017). https://ipg-automotive.com/products-services/test-systems/vil-systems Cited 15 Sep 2017
11. IPG Automotive GmbH.: Vehicle-in-the-Loop. In: News. IPG Automotive (2016). https://ipg-automotive.com/news/article/vehicle-in-the-loop-1 Cited 15 Sep 2017
12. Pfeffer, R., Leichsenring, T.: Continuous development of highly automated driving functions with vehicle-in-the-loop using the example of Euro NCAP scenarios. In: Gühmann, C., Riese, J., von Rüden, K. (eds.) 7th Conference on Simulation and Testing for Vehicle Technology. Springer International Publishing Switzerland, Berlin (2016)
13. Griggs, W.M., Shorten, R.N.: Embedding real vehicles in SUMO for large-scale ITS scenario emulation. In: International Conference on Connected Vehicles and Expo (ICCVE), pp. 962–963. Las Vegas, Nevada, USA (2013)
14. Bock, T., Maurer, M., Färber, G.: Validation of the vehicle in the loop (VIL)—a milestone for the simulation of driving assistance systems. In: IEEE Intelligent Vehicles Symposium, pp. 612–617. Istanbul, Turkey (2007)
15. Sieber, M., Berg, G., Karl, I., Siedersberger, K.-H., Siegel, A., Färber, B.: Validation of driving behavior in the vehicle in the loop: steering responses in critical situations. In: 16th International IEEE Conference on Intelligent Transportation Systems (ITSC), pp. 1101–1106. The Netherlands, The Hague (2013)
16. Galko, C., Rossi, R., Savatier, X.: Vehicle-hardware-in-the-loop system for ADAS prototyping and validation. In: International Conference on Embedded Computer Systems: Architectures, Modeling and Simulation (SAMOS XIV), pp. 329–334. Samos, Greece (2014)
17. TASS International.: Vehicle hardware-in-the-loop. In: Services. TASS International (2017). https://www.tassinternational.com/vehicle-hardware-loop Cited 14 Sep 2017
18. Albers, A., You, Y., Klingler, S., Behrendt, M., Zhang, T., Song, K.: Supporting globally distributed product development with a new validation concept. Procedia CIRP **21**, 461–466 (2014)
19. Quinlan, M., Au, T.-C., Zhu, J., Stiurca, N., Stone, P.: Bringing simulation to life: a mixed reality autonomous intersection. In: IEEE/RSJ International Conference on Intelligent Robots and Systems (IROS), pp. 6083–6088. Taipei, Taiwan (2010)
20. Mangharam, R., Weller, D., Rajkumar, R., Mudalige, P., Bai, F.: GrooveNet: a hybrid simulator for vehicle-to-vehicle networks. In: 3rd Annual International Conference on Mobile and Ubiquitous Systems—Workshops (MobiQuitous) San Jose, California, USA (2006)
21. Wegener, A., Piórkowski, M., Raya, M., Hellbrück, H., Fischer, S., Hubaux, J.-P.: TraCI: an interface for coupling road traffic and network simulators. In: 11th Communications and Networking Simulation Symposium (CNS), pp. 155–163. Ottawa, Canada (2008)
22. Ordóñez-Hurtado, R.H., Griggs, W.M., Massow, K., Shorten, R.N. Intelligent speed advising based on cooperative traffic scenario determination. In: Waschl, H., Kolmanovsky, I., Steinbuch, M., del Re, L. (eds.) Lecture Notes in Control and Information Sciences: Optimization and Optimal Control in Automotive Systems 455:77–92. Springer International Publishing Switzerland (2014)
23. Schlote, A., Häusler, F., Hecker, T., Bergmann, A., Crisostomi, E., Radusch, I., Shorten, R.: Cooperative regulation and trading of emissions using plug-in hybrid vehicles. IEEE Trans. Intell. Transport. Syst. **14**(4), 1572–1585 (2013)
24. Häusler, F., Ordóñez-Hurtado, R.H., Griggs, W.M., Radusch, I., Shorten, R.N.: Closed-loop flow regulation with balanced routing. International Conference on Connected Vehicles and Expo (ICCVE), pp. 1054–1055. Austria, Vienna (2014)
25. Microsoft Corporation.: Microsoft Band (2017). https://www.microsoft.com/Microsoft-Band/en-us Cited 9 Oct 2017
26. Apple Inc.: Watch (2017). http://www.apple.com/uk/watch Cited 9 Oct 2017
27. Google Inc.: Android wear. In: Wear. Android (2017). https://www.android.com/wear Cited 9 Oct 2017

28. Russo, G., Shorten, R.: Hacking an e-bike to help cyclists avoid breathing in polluted air. In: Blog. IBM Research (2017). https://www.ibm.com/blogs/research/2017/08/hacking-an-e-bike Cited 9 Oct 2017
29. Sweeney, S., Ordonez-Hurtado, R., Pilla, F., Russo, G., Timoney, D., Shorten, R.: Cyberphysics, pollution mitigation, and pedelecs (2017). Available at: https://arXiv.org/abs/1706.00646v2 Cited 9 Oct 2017
30. Sweeney, S.: Cyber-physics, pollution mitigation and pedelecs. In: YouTube (2017). https://www.youtube.com/watch?v=265u9KO-9QE Cited 9 Oct 2017
31. CO2Meter.com.: K-30 10,000ppm CO2 sensor. In: Products (2017). http://www.co2meter.com/products/k-30-co2-sensor-module Cited 9 Oct 2017
32. Zhuk, S., Tchrakian, T.T., Moore, S., Ordóñez-Hurtado, R., Shorten, R.: On source-term parameter estimation for linear advection-diffusion equations with uncertain coefficients. SIAM J. Sci. Comput. **38**(4), A2334–A2356 (2016)
33. Birstall Garden & Leisure.: Weathereye professional touch screen weather station with PC interface. In: Products (2017). http://www.birstall.co.uk/products/wea22.html Cited 10 Oct 2017
34. RFID Journal.: Frequently asked questions. In: FAQs. Tools & Resources (2017). http://www.rfidjournal.com/site/faqs Cited 10 Oct 2017
35. Cogill, R., Gallay, O., Griggs, W., Lee, C., Nabi, Z., Ordonez, R., Rufli, M., Shorten, R., Tchrakian, T., Verago, R., Wirth, F., Zhuk, S.: Parked cars as a service delivery platform. International Conference on Connected Vehicles and Expo (ICCVE), pp. 138–143. Austria, Vienna (2014)
36. Verago, R., Naoum-Sawaya, J., Griggs, W., Shorten, R.: Localization of missing entities using parked cars. In: International Conference on Connected Vehicles and Expo (ICCVE), pp. 40–41. Shenzhen, China (2015)
37. RAC Foundation.: Spaced out: perspectives on parking policy. In: Press (2012). http://www.racfoundation.org/media-centre/spaced-out-press-release Cited 10 Oct 2017
38. Liu, N., Liu, M., Lou, W., Chen, G., Cao, J.: PVA in VANETs: stopped cars are not silent. In: 30th IEEE International Conference on Computer Communications (INFOCOM), pp. 431–435. Shanghai, China (2011)
39. Faoláin, Ó., C, Souza, M., Verago, R., Shorten, R.: Cyclist collision prevention system for in-car deployment. In: 21st Annual Society for Design and Process Science International Conference (SDPS-2016), pp. 284–290. Orlando, Florida, USA (2016)
40. Herrmann, A., Liu, M., Shorten, R.: A new take on protecting cyclists in smart cities (2017). Available at: https://arXiv.org/abs/1704.04540v3 Cited 10 Oct 2017
41. Qstarz International Co., Ltd.: BT-Q1000XT. In: GPS Travel Recorder. Products (2013). http://www.qstarz.com/Products/GPS%20Products/BT-Q1000XT-F.htm Cited 11 Oct 2017
42. Torque.: Speedometer. In: Torque Wiki (2012). https://torque-bhp.com/wiki/Speedometer Cited 11 Oct 2017
43. Ordóñez-Hurtado, R.H., Crisostomi, E., Shorten, R.N.: An assessment on the use of stationary vehicles to support cooperative positioning systems. Int. J. Contr. (2017). https://doi.org/10.1080/00207179.2017.1286537
44. ENABLE-S3.: http://www.enable-s3.eu Cited 11 Oct 2017
45. Griggs, W., Russo, G., Shorten, R.: Consensus with state obfuscation: an application to speed advisory systems. In: 19th IEEE International Conference on Intelligent Transportation Systems (ITSC) 2506–2511 Rio de Janeiro, Brazil (2016)
46. Griggs, W., Russo, G., Shorten, R.: Lead and leaderless multi-layer consensus with state obfuscation: an application to distributed speed advisory systems. IEEE Trans. Intell. Transport Syst. (2017). https://doi.org/10.1109/TITS.2017.2700199

Chapter 8
Virtual Concept Development on the Example of a Motorway Chauffeur

G. Nestlinger, A. Rupp, P. Innerwinkler, H. Martin, M. Frischmann,
J. Holzinger, G. Stabentheiner and M. Stolz

Abstract It is well known that the development of future automated driving faces big challenges regarding testing and validation. One strategy to tackle the drastically increased complex interaction of vehicle, driver, and environment is the so-called front-loading approach. This involves virtual development of new vehicle functions enabling early stage testing and validation. Within the funded project *Technology Concepts for Advanced Highly Automated Driving* (TECAHAD), this front-loading approach was applied for a concept development of an automated driving system (ADS)—the Motorway Chauffeur (MWC)—fully responsible for longitudinal and lateral motion of a car on motorways. In the following, we provide an insight on early stage virtual development of this ADS. Topics range from high-level requirements and functional safety investigations to software architecture and major components of the virtual implementation. Finally, first simulation results are shown for some MWC use cases, motivating the planned future real vehicle prototype implementation.

8.1 Introduction

Automated driving is arguably one of the biggest challenges for automotive industry. Besides many technical issues, also legislative questions have to be solved. Currently, the topic of automated driving gained big popularity in public and there is a strong demand not only from car manufacturers but also from their customers.

G. Nestlinger (✉) · P. Innerwinkler · H. Martin · M. Frischmann · M. Stolz
VIRTUAL VEHICLE Research Center, Inffeldgasse 21/A, 8010 Graz, Austria
e-mail: georg.nestlinger@v2c2.at

A. Rupp
Institute of Automation and Control, Inffeldgasse 21/B/I, 8010 Graz, Austria

J. Holzinger
AVL List GmbH, Hans-List-Platz 1, 8020 Graz, Austria

G. Stabentheiner
MAGNA STEYR Engineering AG & Co KG,
Liebenauer Hauptstrasse 317, 8041 Graz, Austria

© Springer International Publishing AG, part of Springer Nature 2019
H. Waschl et al. (eds.), *Control Strategies for Advanced Driver Assistance
Systems and Autonomous Driving Functions*, Lecture Notes in Control
and Information Sciences 476, https://doi.org/10.1007/978-3-319-91569-2_8

The reason is that there are distinctive benefits that make automated driving so attractive nowadays: The technology holds significant societal potential to further reduce vehicle accidents and may give mobility to persons with reduced driving abilities. Looking at comfort and productivity, automated driving offers additional fuel saving, improved traffic throughput, and time gain for the driver, who may focus on different tasks during driving.

Automated driving is evolving from simple state of the art automation (e.g., adaptive cruise control) stepwise toward higher levels of automation [1]. The higher the level of automation, the more complex the interaction between the vehicle and its environment will be. With respect to vehicle testing, this leads to a paradigm shift in testing, since as pointed out in [2], pure physical testing as performed up to now will not be feasible anymore for highly automated driving. As a consequence, additional virtual testing and validation of automated driving functions will be mandatory within a future development process.

In this contribution, first results of a two years funded project TECAHAD, dealing with early-stage virtual development of a so-called MWC are shown. The project aimed to set up a virtual tool environment and use it for the development of a SAE Level 3 ADS [1]. One goal was to explore the possibilities of front-loading in the development of automated driving functions. A special focus was on early simulation-based evaluation and validation of control approaches as well as early consideration of functional safety.

The chapter is structured as follows: Based on the well-established V-model development process, first the high-level function definition is summarized in Sect. 8.2. After this, the main system and software architecture with respect to the MWC are discussed in Sect. 8.3. Section 8.4 gives an overview on a first analysis regarding functional safety. The topic of sensor fusion is addressed in Sect. 8.5. The main features of the MWC are located within the so-called Trajectory Planning, which is explained in Sect. 8.6. First results from simulation are presented in Sect. 8.7. The chapter concludes with Sect. 8.9 summarizing the work and giving an outlook to future activities.

8.2 Requirements

The system requirements act as guideline for the development as well as checking criteria for the final validation tests. This section summarizes the functional specification, area of application, and functional requirements of the MWC.

Function Definition of the MWC and High Level Requirements The MWC is an ADS providing lateral and longitudinal guidance simultaneously. Longitudinal guidance, i.e., speed control, is done with respect to all road users within the sensor ranges. As long as there is no forward vehicle present within the relevant detection range, a set speed v_{set} is maintained considering speed limits as well as safety and comfort limits. If a forward vehicle is in the ego vehicle's lane and cannot be overtaken, it shall be followed with a desired time gap τ_{set}. For more detailed requirements, definition of function and features the interested reader is referred to the standard on full speed range adaptive cruise control [3].

Lateral guidance, i.e., steering control, usually keeps the ego vehicle within its current lane, but also provides the ability of automated lane changes. Consequently, requirements on lateral guidance are based on standards of lane change decision aid systems [4]. Lane change requirements are taken from standards on lane keeping assistance systems [5].

From the requirements of lateral and longitudinal control, it becomes obvious that the MWC combines features of adaptive cruise control (ACC), lane keeping assist (LKA), and lane change assist (LCA) and extends these functions. Therefore, to distinguish the MWC from a "simple" advanced driver assistance system (ADAS) like ACC, it is classified as an ADS.

The MWC's application is limited to motorways, which are characterized by the number of lanes (≥ 2), their lane widths (3–4 m), their curve radius (> 125 m), the speed range (0–130 km/h), and no oncoming traffic. Since the MWC is a Level 3 driving system according to SAE [1], the driver does not have to monitor the MWC, but must be able to take over vehicle control within a defined period of time.

Top Level MWC Logic, States and Transitions For safe handling as well as for high user acceptance, it is of major importance to ensure simple and intuitive handling. A lean and transparent logic has been implemented, allowing straightforward activation and deactivation. The MWC can be in one of the top-level states INACTIVE, STANDBY, or ACTIVE. Transitions between these states are performed according to Fig. 8.1.

The state INACTIVE is the default state after starting the ignition of the vehicle. Within this state, the MWC is off and does not perform any calculations nor provides any outputs. The MWC is enabled by the driver via an appropriate cockpit switch. Doing so, the MWC's state changes from INACTIVE to STANDBY where all activation criteria are checked. The driver is still responsible for controlling the vehicle. Once all activation criteria are checked and fulfilled, the MWC proceeds to state ACTIVE and takes over vehicle control. The driver must always be informed about the current state of the MWC, e.g., by dashboard display.

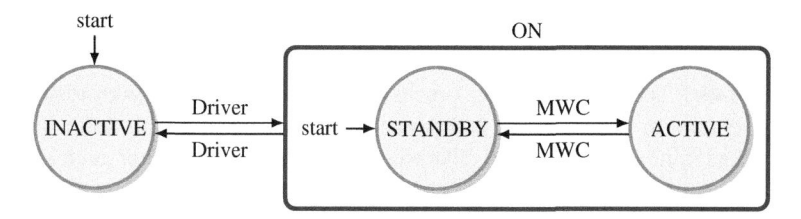

Fig. 8.1 Top-level states and transitions of the MWC. In the states INACTIVE and STANDBY, the driver is responsible for control. If the MWC is in state ACTIVE, the system is controlling the vehicle

To transit from state STANDBY to ACTIVE and remain therein, the following system thresholds have to be fulfilled.

- The MWC supports ego vehicle velocities from 0 to 130 km/h.
- The minimum curve radius is 125 m.
- All components used for the MWC have to be working correctly, i.e., their status signal must not indicate an error.
- Seat occupancy has to be positive, indicating the driver's presence.
- The gear stick has to be in drive position.
- No trailer is attached to the ego vehicle to ensure correct observation of the ego vehicle's rear zone.
- The ego vehicle has to be on a highway (confirmation via navigation system).
- The lane width is at least 3 m and does not exceed the maximum width specified by the sensor capabilities on lane detection (camera, LIDAR).

These requirements are implemented using a combination of functions listed in Sect. 8.3. They are checked within the so-called *Behavioral Planning* component choosing the required MWC state and operating mode.

8.3 Software Architecture

An important aim of the project was a modular software structure that can be used (or reused with few adaptations) for different highly automated driving functions within concept development phase, where the focus is on the simulation domain. The basic software architecture used for the MWC is shown in Fig. 8.2. The MWC is built up by five core software components (blue) which are shortly explained in the following. The practically oriented reader may be interested in the information, that the implementation was carried out using commercial development tools: MATLAB®/Simulink® [6] and CarMaker® [7].

Operational Constraints The complexity of the MWC and its algorithms requires many constraints to be considered for performing automated driving in ACTIVE state. To avoid parameter definitions scattered over all MWC components, this unit serves as a single source for all system thresholds relevant to the MWC, such as velocity and acceleration limits.

Vehicle Model The *Vehicle Model* processes sensor information with respect to the ego vehicle. In this software component, usually sensor signals from inertial measurement units (IMU) or signals like wheel speeds are fused within model-based observers to estimate vehicle states such as position, velocity, acceleration, yaw angle, and yaw rate [8]. Due to the concept level of the project and the focus on the planning part, this component has not been implemented in detail, but instead the corresponding vehicle states have been directly reused from simulation.

Environment Model The aim of the *Environment Model* unit is to provide a mathematical representation of the ego vehicle's surroundings, i.e., its position on the

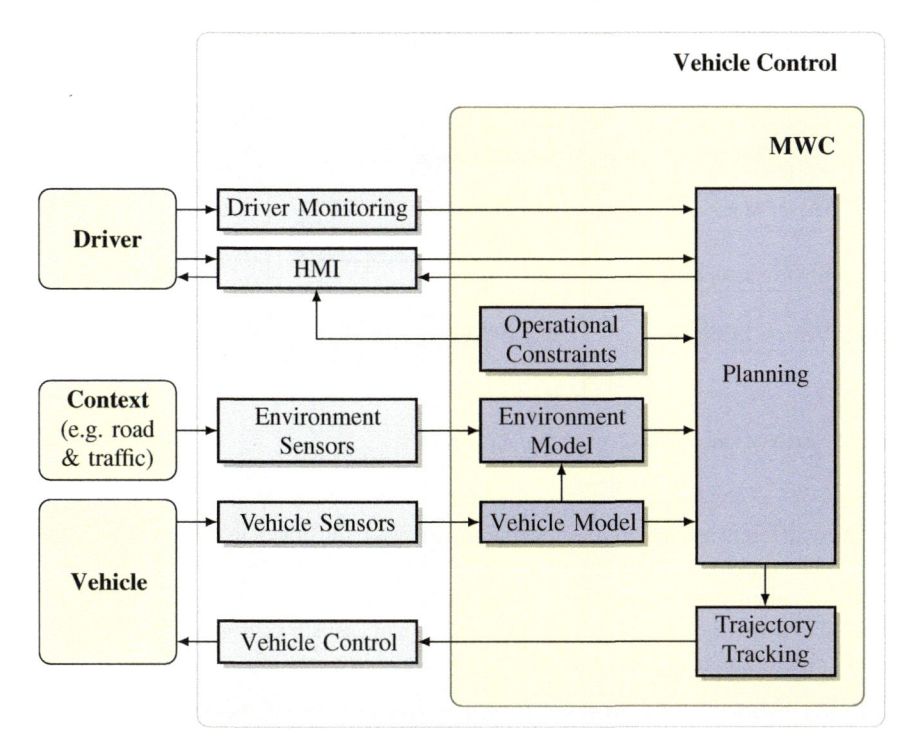

Fig. 8.2 Software architecture of the MWC and its interfaces within the vehicle control. Arrows denote the direction of information flow between the components. For closed-loop evaluation in simulation the behavior of the driver, the interaction with the traffic as well as the vehicle's reaction to control actions and its interaction with the road are simulated using state of the art tools

road, geometrical road properties, distances to and velocities of other objects, position and status of traffic signs, interpretations of the current traffic situation, and so on. Main inputs to the *Environment Model* are preprocessed sensor information from environment sensors, which usually can be split into:

- Information about detected surrounding objects: This information is provided using a so-called object list per sensor. Each item of this object list corresponds to one object, which represents, e.g., another vehicle, a traffic sign, a pedestrian, an obstacle, or a traffic signal. Each object itself holds various properties such as its classification, its speed and acceleration, position.
- General information about ambient light, weather conditions (e.g., rain), temperature, GPS position, barometric pressure, ego vehicle localization with respect to the road, etc.

Additionally preprocessed information from the *Vehicle Model* is used as an input, since its information has to be integrated into the *Environment Model* in order to represent the vehicle in its environment.

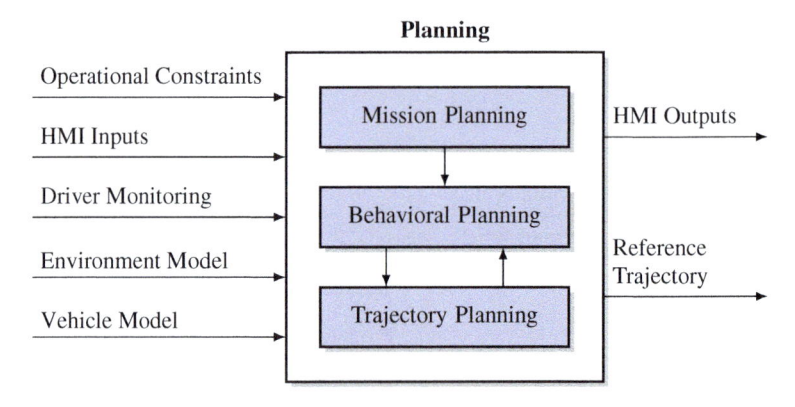

Fig. 8.3 Architecture of MWC component *Planning*

Since different sensors may detect the same object, the object lists from all sensors have to be combined to one single object list. This involves two steps. The first step is the so-called data association, where detected objects of different sensors are mapped to a common set of objects. In a second step, sensor fusion (on feature level) is performed. The approach used will be explained in Sect. 8.5.

Planning This software component is responsible for planning the vehicle's actions. It is subdivided into different layers of planning, according to conventional state of the art approaches as shown in Fig. 8.3. First a top level performs navigation planning and solves strategic route decisions. On the second layer, behavioral planning is done. The third layer deals with planning the ego vehicle's desired lateral and longitudinal reference trajectories.

In contrast to existing approaches, a main characteristic and major design target of the presented MWC are to develop from higher automation levels to lower levels instead of the conventional bottom up approach. This means, instead of using lower level functions (e.g., ACC) within higher automated driving functions, the approach was to use the higher automated driving function (the MWC with full longitudinal and lateral control) and deactivate features to downgrade to lower level functions (e.g., ACC). This led to some distinct advantages:

- Motion planning is done in a consistent and straightforward way. Since ACC, LCA, and LKA only partly consider all the scenarios relevant for the MWC, an additional logic is needed to cover missing cases. This would lead to a functionality which is difficult to validate, since it bases on rather complex case by case decisions, e.g., for target selection in ACC following mode.
- As will be shown later, the *Behavioral Planning* layer becomes more straightforward, which leads to less effort in testing and thus less risk of failure.
- The overall behavior of the vehicle will be more consistent when comparing different levels of automation.
- Tuning the behavior is straightforward.

Mission Planning: This unit represents the navigation layer of the proposed software architecture. It holds all algorithms deciding the global driving route, i.e., from the vehicle's current location to the final location of the trip. Outputs are the desired lane and speed limitations due to speed restrictions or curve radii as a function of the planned route.

Behavioral Planning: This is the main decision unit of the MWC. The *Behavioral Planning* checks all MWC components for proper functionality, implements the state machine from Fig. 8.1, controls the transition between STANDBY and ACTIVE state and activates or deactivates the appropriate *Trajectory Planning* operation mode. It also evaluates the inputs from *Mission Planning* according to the current vehicle state and provides the results to the *Trajectory Planning* unit.

Trajectory Planning: This layer holds the trajectory planning for all supported ADAS functions. The MWC's Trajectory Planning covers the features of an adaptive cruise control (ACC), lane keeping assistant (LKA), traffic jam assistant (TJA), and lane change assistant (LCA); see also Sect. 8.6. Specific activation and deactivation of features of the *Trajectory Planning* result in realization of the desired ADAS function and are done by the *Behavioral Planning* unit.

New ADAS functions can easily be added by providing the lateral and/or longitudinal reference trajectory interface as required by the *Trajectory Tracking* control unit (see section below) or by extending the existing *Trajectory Planning*. The feedback connection to *Behavioral Planning* contains status signals of the ADAS functions.

Trajectory Tracking The *Trajectory Tracking* component implements the feedback control of the vehicle's lateral and longitudinal motion with respect to the reference trajectories provided by the *Trajectory Planning* subsystem. Its implementation uses a separate control strategy for the vehicle's lateral and longitudinal motion, as discussed in the following sections. Since the vehicle's motion is planned within the Trajectory Planning, the resulting reference values for lateral and longitudinal control are always consistent. Depending on the rate of execution, repetitive re-planning also ensures consistency in the presence of unknown disturbances.

Lateral Tracking The feedback control for lateral tracking is based on the well-known single-track model [9]. To accomplish the lane keeping (and also the lane changing) task, the lateral offset y_L from the desired position and the angular offset ϵ_L from the desired orientation were defined as the control errors of the so-called lane tracking model according to Fig. 8.4 [10]. The index L indicates that the control errors are defined at the so-called look-ahead distance L ahead of the vehicle's center of gravity. To compensate the linearly changing curvature κ_L of a clothoidal highway, two additional states forming a chain of integrators using the lateral offset y_L as its input were defined. Finally, a simple steering model was used to replace the single track model's control input, namely the front wheel steering angle, by the steering wheel torque M_H. The resulting state-space model was used to design a standard LQR feedback controller.

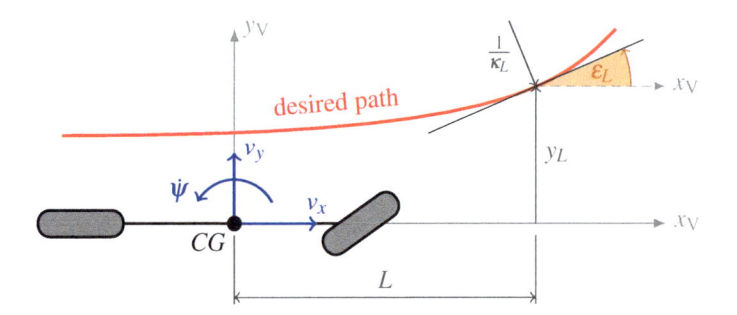

Fig. 8.4 Control error variables y_L and ϵ_L based on vehicle position and desired path using the look-ahead distance L

The *Trajectory Tracking* uses the following input signals to define its lateral reference trajectory:

- the look-ahead distance L,
- the actual lateral offset y_L,
- the desired lateral offset,
- the actual angular offset ϵ_L
- and the transition time to be used with the flatness-based bumpless transfer method.

Additionally, measurements of the yaw rate $\dot{\psi}$, steering wheel angle, steering wheel angular velocity, and the steering wheel torque at the instant of lateral control activation are used.

A model-based anti-windup method avoids the potential risk of integrator windup. To ensure a smooth transition from manual to automatic steering at the instant of switching from MWC state STANDBY to ACTIVE, a flatness-based bumpless transfer measure was applied. Since the mathematical model depends on the vehicle speed v_x and the look-ahead distance L, both, the controller feedback parameter and the anti-windup parameter were gain scheduled. Further, details on the lateral control strategy can be found in [11].

Longitudinal Tracking The longitudinal tracking features a simple PI controller. Its reference trajectory only uses the desired longitudinal acceleration. The deviation of the current value is defined as the control error. Proportional and integral coefficients were tuned by iterative simulations assuming a standard configuration of the vehicle under control. To avoid windup phenomena, the state of the integrator is reset if its output exceeds a predefined limit. The longitudinal tracking unit provides gas and brake pedal positions.

8.4 Functional Safety for Automated Driving Functions

Within the project, the following three automated driving systems were analyzed regarding their functional safety aspects according to concept phase of ISO 26262 [12]: Motorway Assistant (MWA), Motorway Assistant+ (MWA+) and Motorway Chauffeur (MWC) (see Table 8.1).

For each of these functions, the steps depicted in Fig. 8.5 were performed.

In the following, the conclusions and results are summarized. As the focus of the project was on concept level, investigations regarding functional safety were performed on the top level only. First the *Item Definition* of the MWC was carried out. This involved the definitions of functional and nonfunctional requirements as well as analyzing dependencies between the item and its environment (properties of interfaces between the item, HMI, sensors, and actuators). The following *Hazard Analysis and Risk Assessment* (HARA) splits into four steps. For the MWC, five major hazards have been identified. Situation analysis resulted in four groups of situations to be considered: locations, weather conditions, driving maneuvers, and environmental conditions. Within the *Risk Assessment*, all combinations of hazards and situations have been examined and rated with respect to their exposure (E0–E4), their severity (S0–S3) and their controllability (C0–C3). The HARA performed for

Table 8.1 Evolution of the MWC. For the MWA, the driver is fully responsible to carry out lane changes (LC). The MWA+ provides lane change recommendations, but their automated execution still needs to be unblocked by the driver. The MWC features fully automated lane changes without driver interaction

ADS	Active LC	Passive LC	Failure mode	Safe state	SAE level	ASIL
MWA	–	–	Fail silent	Driver	2	B
MWA+	–	+	Fail silent	Driver	2	C
MWC	+	–	Fail operational	System	3	D

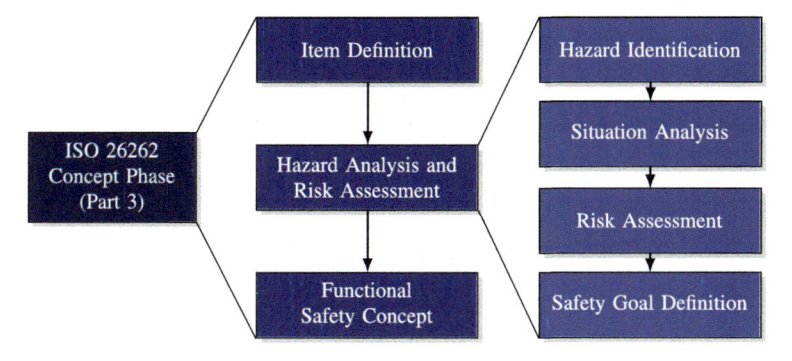

Fig. 8.5 Overview of safety activities in concept phase of ISO 26262

the three ADS evolution stages showed that increasing complexity from MWA to MWC leads to a higher Automotive Safety Integrity Level (ASIL) classification. For the MWA, the highest ASIL is B, for the MWA+, the highest ASIL is C, and for the MWC, the highest ASIL is D. The main reason for this increase of the ASIL is the controllability of the ADS by the driver. The controllability is classified as *simply controllable* for MWA and MWA+ because the driver must always be present, monitor all driving scenarios and be able to take over control within a short time (e.g., 2 s). In contrast to that the MWC's controllability has to be rated as *difficult controllable* (C2) or even *uncontrollable* (C3) because the driver is not required to monitor the driving maneuvers at all times.

As a result, based on all safety critical hazardous events of the HARA, ten top-level safety requirements, the so-called *safety goals*, were defined.

The controllability of the ADS by the driver also affects the safe state of the ADS and therefore the safety concept. The safe states for MWA and MWA+ are deactivating the ADS and taking over vehicle control by the driver. This safe state is not possible for the MWC because in general, the driver is not monitoring the driving situation at all times—the driver must only be able to take over vehicle control within a specified period of time. Therefore, it is necessary to provide a fail-operational system in case of a malfunction of the MWC. This requirement affects the technical architecture of the vehicle and its subsystems for reliability, availability, and safety. The technical architecture and the fail-operational constraints are in focus of the follow-up project based on the results of TECAHAD. For further considerations of functional safety for ADS, the reader is referred to [13].

8.5 Sensor Fusion

Nowadays, the most important sensors for automated driving are camera and radar. Figure 8.6 illustrates the sensor configuration proposed for the MWC enabling a 360° environment perception. Having multiple sensors with an overlapping detection range makes it necessary to perform sensor fusion on their outputs to get one single environment representation.

Within the proposed architecture for the MWC, sensor fusion is performed at two different locations:

1. on signal level within the vehicle model and
2. on feature level within the environment model.

The topic of sensor fusion with respect to the vehicle model is already handled in special control units, e.g., the electronic stability program (ESP). The corresponding signals are assumed to be known and within the early stage development directly taken from simulation. Since the MWC is assumed to maximize comfort, no high dynamic driving maneuvers have to be considered (e.g., drifting) and corresponding control topics can be neglected.

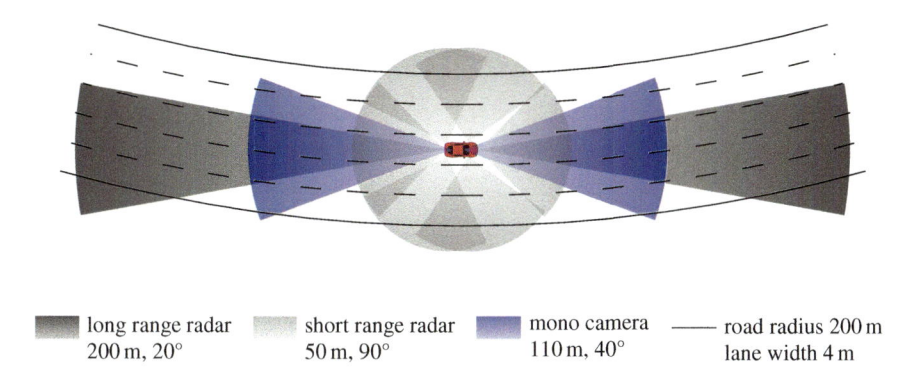

Fig. 8.6 Sensor configuration of the MWC. The sensor ranges are scaled down by a factor of 4 to fit the page width. The proposed configuration comprises 2 long-range and 6 short-range radar sensors as well as two cameras

In the following, we focus on the second sensor fusion part necessary for all higher automation levels. The main motivation to perform sensor fusion of object properties on feature level was to ensure modularity and extendability of the overall software structure as well as to keep small interfaces.

For the sensor fusion task, different approaches have been proposed in the literature. The interested reader is referred to [14], where the approach used in this project is described in detail. The working principle is illustrated in Fig. 8.7 for a setup of three different sensors. As time increases from left to right, measurements are taken and the sensor specific preprocessing is performed resulting in object data (e.g., an objects relative position and velocity). This information is used in sensor specific instances of Kalman filters, one per detected object. In Fig. 8.7, green circles symbolize the result of one Kalman filter step (prediction and correction). Gray circles denote missing measurements and only use prediction, which is allowed for a defined number of samples until the corresponding Kalman filter of an object is switched off (and maybe reinitialized for another object). By using this strategy, robustness against object data missing for a defined duration can be increased. Objects can be tracked even if they are hidden by other objects or move outside the sensor range temporarily. All filters share the same definition of states. In the project, two states for plane position and two for corresponding velocities have been used. Note that measurements may be sensor specific and may have to be treated differently with respect to the sensor.

Finally, for each detected object an additional Kalman filter is used to fuse the states of the sensor-specific filters (denoted in blue in Fig. 8.7). Within the correction step of the fusion filter, the most up to date information of the sensor filter's states is used (blue dashed and gray dash-dotted lines). Note that for maximum performance of the algorithm, the duration between the update of the fusing filter and the last update of a sensor specific filter has to be considered. This time duration—loosely speaking the age of the measurement—is used to extrapolate the measurement to

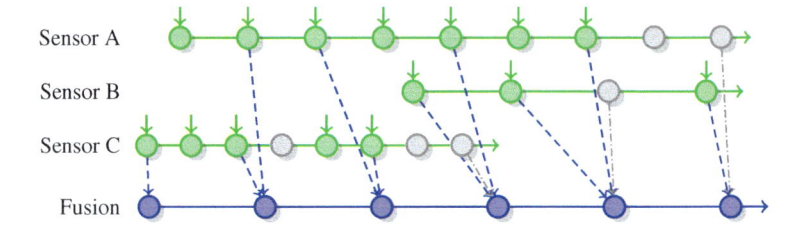

Fig. 8.7 Sensor fusion of object properties on feature level according to [14]. Time increases from left to right. Circles denote states of an object model within a Kalman filter. The prediction step is symbolized by the horizontal connection between two circles. Vertical solid arrows denote samples with measurement inputs and the corresponding correction step in the filter. The correction step of the fusion filter uses the state information of the sensor-specific filters instead of measurements (dashed and dash-dotted arrows)

the point in time, where the fusion takes place. Therefore, storing the time stamp for each measurement is necessary.

The described approach can be applied to sensors with different sampling rates and is straightforward to be extended in case of adding sensors. This two-level approach leads to direct consideration of sensor specific noise within the sensor-specific object filters.

8.6 Trajectory Planning

Previous assistance systems such as ACC and LKA only affect longitudinal or lateral motion and define their references via target vehicles or target lanes. The MWC, however, is capable of changing lanes and adapting velocities at the same time. To compute a safe and feasible reference trajectory, i.e., the ego vehicle's path and velocity as a function of time, the Trajectory Planning level has been introduced. This level consists of several computational blocks as shown in Fig. 8.8.

First, a coordinate transformation of all vehicles into road coordinates is performed. Every detected vehicle is assigned a longitudinal distance along the road and the lateral offset with respect to the rightmost lane center. Second, a set of trajectories to different predefined endpoints is computed for straight road segments, which has been described, e.g., in [15]. A common approach for lateral trajectory generation is the computation of fifth-order polynomials as in [16, 17], yielding smooth transitions that minimize the jerk and thus maximize comfort. The polynomial coefficients are defined by the actual states of the ego vehicle, and its desired end conditions, which can be chosen in consideration of the road characteristics. The velocity trajectories can be defined by profiles as proposed in [18, 19], or by using fourth-order polynomials [17]. The latter has been implemented in the MWC, minimizing jerk in longitudinal direction. Different lateral trajectories are then combined with these velocity trajectories.

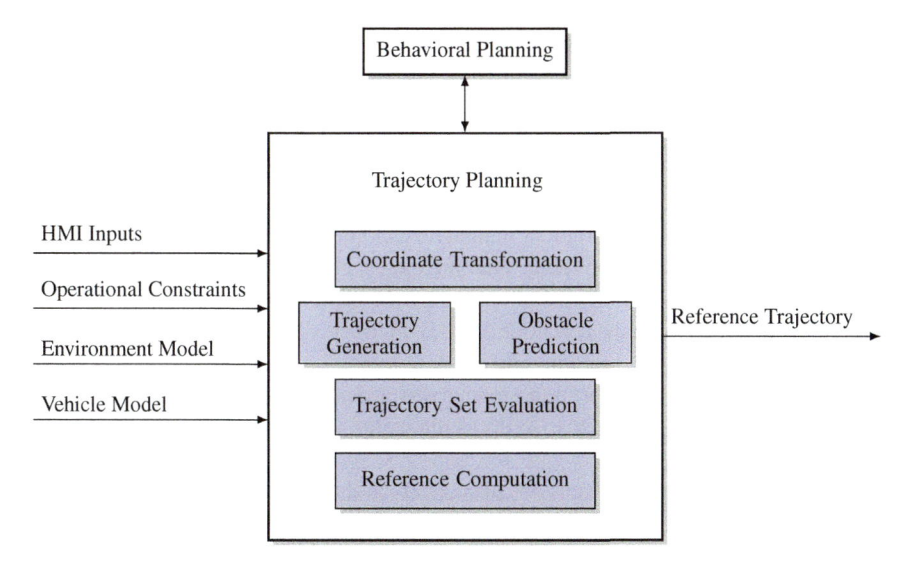

Fig. 8.8 Trajectory Planning unit consists of several sub-units. The higher-level Behavioral Planning is responsible to switch it on/off and control its working modes

The other traffic participants' trajectories are predicted in parallel using the same approach of trajectory planning assuming constant velocities. Their trajectories are then chosen based on collision avoidance and desired lane only. In addition, an observer has been designed that decides whether the obstacle is predictable. If an obstacle does not behave as predicted, its current lateral offset is assumed to be constant over the predicted time horizon.

Typically, the polynomials are generated over a fixed time horizon for standard scenarios in one planning cycle. In the proposed MWC, another set of polynomials is generated over a shorter time horizon in a second planning cycle, which represents overtaking maneuvers. The second cycle assumes a constant velocity in order to reduce the total number of trajectories. This approach has improved the performance of overtaking when faster rear vehicles are present, since the ego vehicle is able to predict the second lane change as well, i.e., yielding for approaching obstacles can be considered. If the maneuver planning does not yield collision-free trajectories, another planning strategy can be activated: Multiple planning cycles over shorter time horizons are proposed that yield a higher number of possible paths. The combination of these two planning strategies showed promising results in simulations and yielded better performance than trajectories generated over only one planning cycle.

Then, the generated trajectory set of the ego vehicle is evaluated based on a cost function in order to pick the "best" trajectory with respect to several aspects. The most important components of this cost function are

- Safety: The goal is to keep safety distances during standard maneuvers and avoid collisions or limit the damage of collisions. For this purpose, ellipses are computed that are centered at other traffic participants. If an ellipse is entered by a generated trajectory, the collision cost component increases significantly. Additionally, the velocity and time instance of crossing an ellipse are considered in the cost component.
- Acceptance of the driver: Efficient maneuvers should be computed that are similar to human driving behavior while guaranteeing safety. The driver can choose a desired velocity that should be maintained by the vehicle. The cost component for the desired velocity consists of a static part, which considers the velocity at the end of the trajectory, and a dynamic part for fast transitions to this end velocity. If a lane change is allowed, overtaking maneuvers are performed in order to maintain the desired velocity. In order to keep to the right lane, a cost component for the desired lane is used.
- Comfort: High jerk and unnecessary accelerations should be avoided and are penalized in cost components for lateral and longitudinal direction.

For this purpose, the generated trajectories of the ego vehicle and the predicted trajectories of the other participants are evaluated at sampled points with a sampling time T_s. The cost components are computed using the sampled points of the ego vehicle trajectories. In a last step, all cost components are summed up and the trajectory with the least cost is chosen. In order to avoid high-frequency switching between trajectories, a filter can be added. This evaluation is the core of the trajectory planning level. Due to its extensive considerations of the surrounding vehicles, the behavioral planning level can be reduced significantly. For example, the target vehicle selection can be removed completely. The generated trajectories and the best trajectories after evaluation are shown in Fig. 8.9.

Finally, the best trajectory is transformed into the vehicle's local coordinate system, taking into account the curvature of the road. The reference trajectories are then forwarded to the tracking controller. If the sampling rate of the trajectory plan-

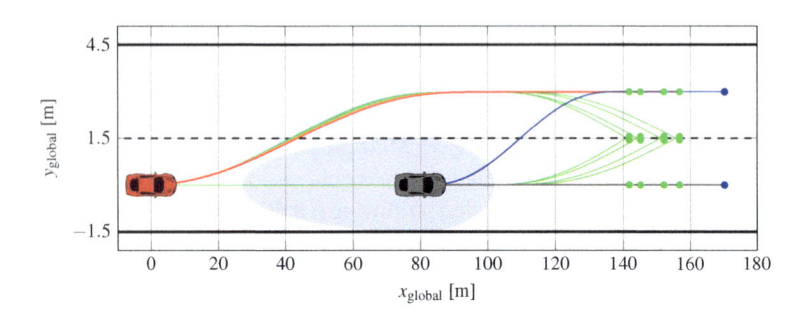

Fig. 8.9 Resulting trajectories of the generation process (green and blue) for the ego vehicle (red car) and the obstacle (gray car). Endpoints are represented by green and blue circles, respectively. The evaluated best trajectory of both vehicles is highlighted by red respectively gray line. The ellipse around the obstacle represents the safety distance that must be maintained

ning level is larger than the sampling rate of the controller, the references are computed until the next planning instance is performed. The lateral controller expects the desired lateral offset at the look-ahead distance. The longitudinal controller's reference consists of the desired acceleration. Further, details can be found in [20].

8.7 Simulation Results

This section presents some of the MWC's features using numerical simulation results:

- A merging maneuver to handle the reduction of the number of lanes (Sect. 8.7.1). Since the merging scenario is a quite common situation in highway driving, it must be handled by the MWC.
- A collision avoidance maneuver by means of acceleration limits exceeding common comfortable zones (Sect. 8.7.2). For SAE Level 3 systems, the driver is not required to take over vehicle control on demand but within a defined period of time. For that reason, it might be necessary to perform collision avoidance maneuvers before the driver takes over vehicle control or to bring the vehicle in a safe state.
- A bumpless transfer safety feature for the lateral Trajectory Tracking unit (Sect. 8.7.3). The aim of this functionality is to ensure a safe and comfortable transition between operating points with different lateral offsets. Although results are only shown for the activation of the lateral Trajectory Tracking component, this feature can also be used while lateral tracking is throughout active, e.g., at suddenly appearing lane markings splitting the current lane into two separate lanes.

8.7.1 Merging Maneuver

This simulation shows a maneuver for merging traffic from two lanes into one (Fig. 8.10). The ego vehicle is approaching two obstacle vehicles, both cruising at a constant speed of 27 m/s. Since the left lane is free and the ego vehicle's desired velocity is set to 130 km/h $\approx$ 36 m/s, it initiates an overtaking maneuver by accelerating and changing to the left lane. At around $x_{\text{global}} = 575$ m, the Behavioral Planning unit triggers the Trajectory Planning to merge to the desired lane, which is the right lane. As a result, the ego vehicle merges into the free space between the obstacle vehicles.

8.7.2 Collision Avoidance Maneuver

Figure 8.11 shows the simulation results of a collision avoidance maneuver. The ego vehicle is driving with an initial velocity of 130 km/h = 36 m/s on the right lane of

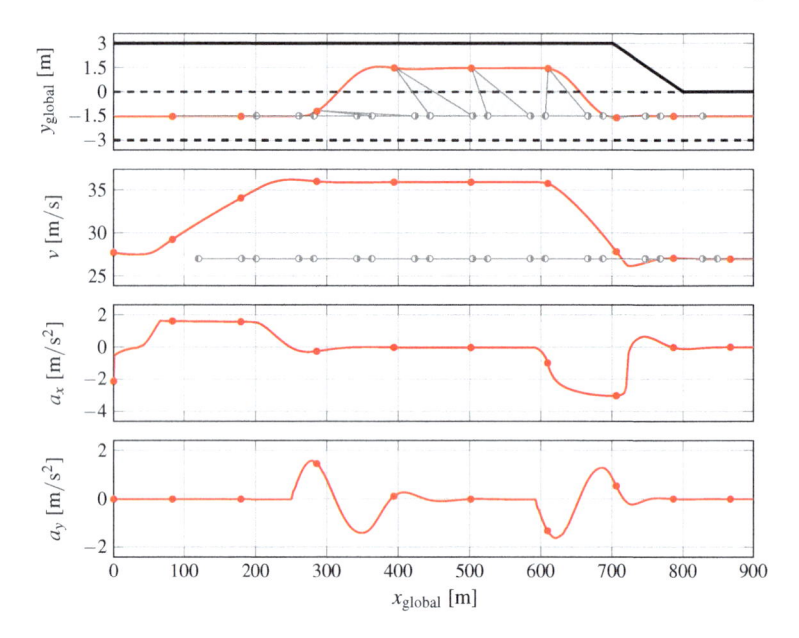

Fig. 8.10 Ego vehicle path (red) during a merging maneuver. Two obstacle vehicles (gray) are driving at the same constant speed of 27 m/s. Connected plot markers forming filled triangles in the first row represent snapshots of ego and obstacle vehicles at the same instant of time, although the triangle is only visible during the overtaking maneuver. To keep the plot as clear as possible, the first row does not show plot markers for the whole plot range

a straight road. Two vehicles, namely A and B, are in standstill in front of the ego vehicle. Vehicle A is located 90 m in front on the right lane, vehicle B 130 m in front on the left lane. Each lane has a lane width of 3 m.

As time increases, the ego vehicle approaches vehicle A on the right lane. To avoid a collision, it initiates a lane change to the left lane while simultaneously decelerating (at approx. 40 m). After passing by vehicle A, the left lane is blocked by vehicle B. Since the ego vehicle is still too fast to stop behind vehicle B, the MWC again decelerates and turns back to the right lane. After passing the vehicles A and B, the MWC accelerates to a predefined velocity to clear the already blocked street section.

The short oscillation at around 130 m is a result of shifting from the fifth to the fourth gear. During this maneuver, the lateral acceleration a_y reaches the single-track model's range of validity of $\pm 4\,\text{m/s}^2$ [21].

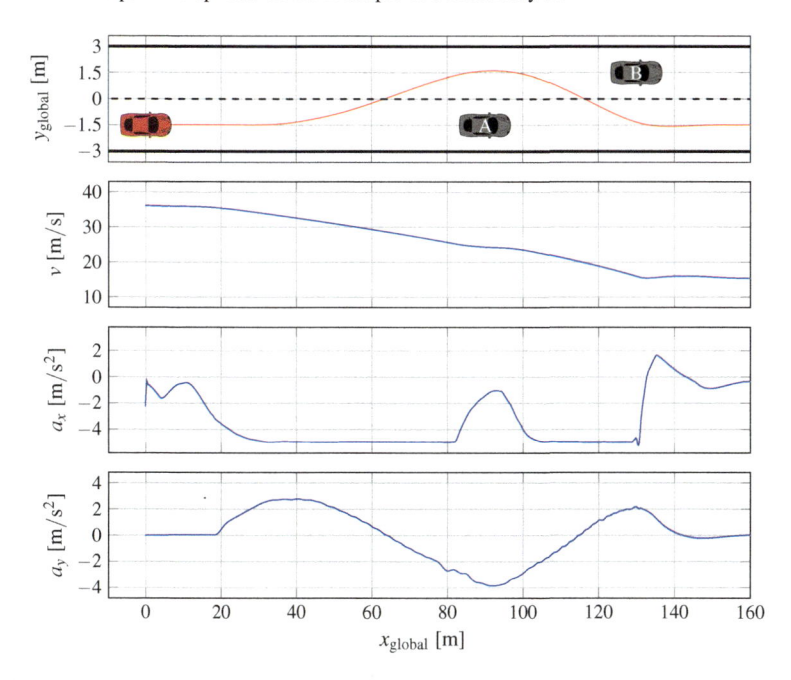

Fig. 8.11 Ego vehicle (red) path during the collision avoidance maneuver. Both obstacle vehicles (gray, labeled A and B) are in standstill throughout this maneuver. Please note that the vehicles' widths are scaled to match the plots axis; therefore, their length of around 4.5 m is not true to scale

8.7.3 Bumpless Transfer for Lateral Tracking

The Trajectory Planning already provides smooth desired trajectories. Nevertheless, the bumpless transfer measure is used as a safety feature at the lateral Trajectory Tracking unit. For further details, the interested reader is referred to [11].

In order to demonstrate the bumpless transfer of the lateral tracking, two simulation runs compare controller activation with and without bumpless transfer measures. For that, a clothoid with increasing curvature $\kappa = 0, \ldots, 1/300 \, \text{m}^{-1}$ was used as the reference path. The simulation was carried out applying the control on a non-nominal plant, i.e., a slightly different model within the commercial simulation software. The lateral tracking system was activated at time instant $t_* = 10 \, \text{sec}$, where the lateral offset $y_L \approx 1.5 \, \text{m}$.

The simulation results are shown in Fig. 8.12. In contrast to the case without bumpless transfer measures (solid blue line), applying the flatness-based bumpless transfer method (dotted red line) allows for eliminating the bump in the steering torque control input M_H and providing a smooth control. As a result, lateral acceleration a_y and lateral jerk j_y are reduced significantly in magnitude. In addition, the performance of controlling the lateral offset y_L was improved by reducing undershot and settling time.

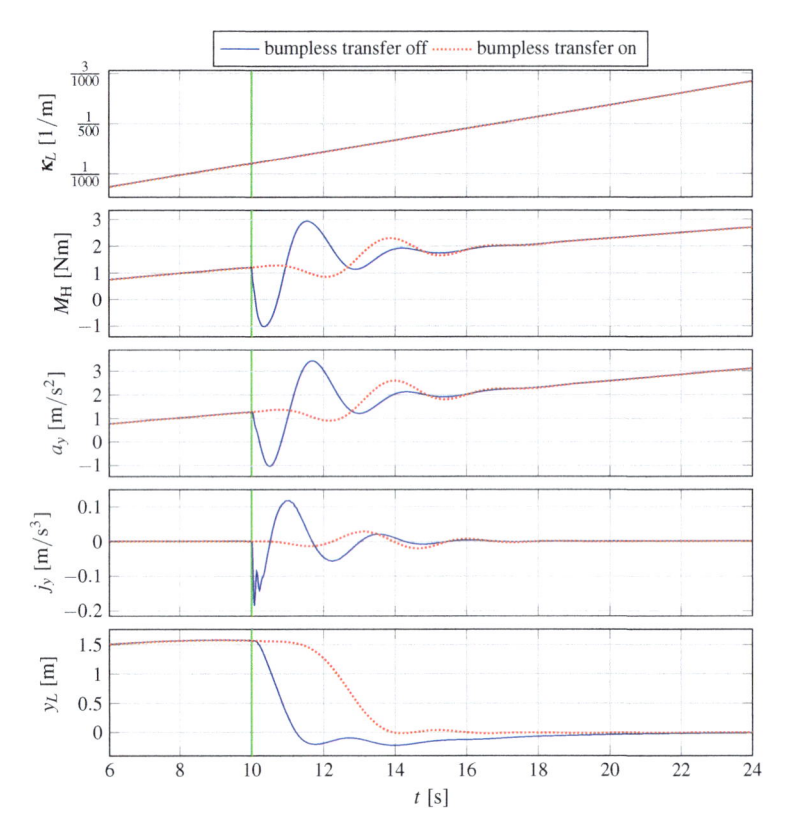

Fig. 8.12 Comparison of lateral tacking activation at $t_* = 10\,\text{s}$ (indicated by the green line) with and without flatness-based bumpless transfer using a transition time $T = 6\,\text{s}$

8.8 Conclusion

From the authors point of view, the application of front-loading for the development of new ADAS and ADS is highly recommended. Compared to a real prototype vehicle, simulation toolchains are easily manageable and adaptable. They do not require expensive test tracks but allow for fully reproducible simulation studies and easy data logging and data evaluation.

While the tool environment of MATLAB/Simulink and CarMaker enabled an easy implementation of algorithms, it was not able to perform in real time on a standard business laptop (only $0.4 \times$ real-time). This drawback can be compensated by running simulations simultaneously on multiple work machines. Assuming a total of 10 laptops, an average speed of $20\,\text{m/s} = 72\,\text{km/h}$ and the above real-time factor about 120,000 km of testing could be performed within one week (equals 168 h).

One remaining limitation of the validation phase is multi-vehicle scenarios. Defining virtual test runs with more than about five vehicles is still a challenging task and, due to the vehicle interaction, leads to highly increased complexity. A solution might be to use tools specifically designed for that purpose. On the other hand, this approach would result in a co-simulation which is a complex topic on its own.

8.9 Outlook

This chapter gave an insight on the work in progress of a nationally funded project dealing with the early stage development of an automated driving system. The focus of the project was on pure simulation and virtual development of highly automated driving. On the example of a Motorway Chauffeur (SAE Level 3), a front-loading approach was used for control development as well as for accompanying considerations regarding functional safety. As a result, it can be stated that a combination of state of the art tools is able to cover early development. It is worth noting that evaluation and testing on concept level can be realized fully based on virtual prototypes. From the authors' perspective in future, virtual prototypes will gain importance also for final testing and validation of automated driving.

A follow-up project of another two years is starting end of 2016, dealing with first prototype implementations of the functionality. Planned prototypes are a driving simulator, scaled model trucks, and a demonstration vehicle with full software access to steering, braking, and acceleration.

Acknowledgements The publication was written at VIRTUAL VEHICLE Research Center in Graz, Austria. The authors would like to acknowledge the financial support of the COMET-K2— Competence Centers for Excellent Technologies Programme of the Federal Ministry for Transport, Innovation and Technology (bmvit), the Federal Ministry for Digital, Business and Enterprise (bmdw), the Austrian Research Promotion Agency (FFG), the Province of Styria and the Styrian Business Promotion Agency (SFG).
They would furthermore like to express their thanks to their supporting industrial and scientific project partners, namely AVL List GmbH, MAGNA STEYR Engineering AG & Co KG and to the Institute of Automation and Control at the Technical University of Graz.

References

1. SAE International: Taxonomy and Definitions for Terms Related to On-Road Motor Vehicle Automated Driving Systems (2014). https://doi.org/10.4271/J3016_201401. http://doi.org/10.4271/J3016_201401
2. Winner, H.: ADAS, Quo Vadis, pp. 1557–1584. Springer International Publishing, Cham (2016)
3. ISO 22179:2009: Intelligent transport systems—Full speed range adaptive cruise control (FSRA) systems—Performance requirements and test procedures. Tech. rep., International Organization for Standardization (2009)
4. ISO 11270:2014: Intelligent transport systems—Lane keeping assistance systems (LKAS)— Performance requirements and test procedures. Tech. rep., International Organization for Standardization (2014)

5. ISO 17387:2008: Intelligent transport systems—Lane change decision aid systems (LCDAS)—Performance requirements and test procedures. Tech. rep., International Organization for Standardization (2008)
6. MATLAB: Version 7.14 (R2012a).: The MathWorks Inc., Natick, Massachusetts (2012)
7. IPG Automotive GmbH.: CarMaker 4.0. Karlsruhe, Germany (2012)
8. Neubauer, L.: Absolute Position Estimation with GPS/INS Sensor Fusion
9. Riekert, P., Schunck, T.: Zur Fahrmechanik des gummibereiften Kraftfahrzeugs. Ingenieur-Archiv **11**, 210–224 (1940). https://doi.org/10.1007/BF02086921
10. Kosecka, J., Blasi, R., Taylor, C., Malik, J.: A comparative study of vision-based lateral control strategies for autonomous highway driving. In: Proceedings of IEEE International Conference on Robotics and Automation, 1998, vol. 3, pp. 1903–1908 (1998)
11. Nestlinger, G., Stolz, M.: Bumpless transfer for convenient lateral car control handover. IFAC-PapersOnLine **49**(15), 132–138 (2016). In: 9th IFAC Symposium on Intelligent Autonomous Vehicles IAV 2016, Leipzig, Germany, 29 June–1 July 2016
12. ISO/DIS 26262-1—Road Vehicles Functional Safety Part 1 Glossary. Tech. rep., Geneva, Switzerland (2009)
13. Martin, H., Tschabuschnig, K., Bridal, O., Watzenig, D.: Functional Safety of Automated Driving Systems: Does ISO 26262 Meet the Challenges, pp. 387–416. Springer International Publishing, Cham (2017)
14. Stüker, D.: Heterogene Sensordatenfusion zur robusten Objektverfolgung im automobilen Straßenverkehr. P.h.d. Thesis, Carl von Ossietzky-Universität Oldenburg (2004)
15. Urmson, C., Anhalt, J., Bae, H., Bagnell, J., Baker, C., Bittner, R., Brown, T., Clark, M.N., Darms, M., Demitrish, D., Dolan, J., Duggins, D., Ferguson, D., Galatali, T., Geyer, C., Gittleman, M., Harbaugh, S., Hebert, M., Howard, T., Kolski, S., Likhachev, M., Litkouhi, B., Kelly, A., McNaughton, M., Miller, N., Nickolaou, J., Peterson, K., Pilnick, B., Rajkumar, R., Rybski, P., Sadekar, V., Salesky, B., Seo, Y., Singh, S., Snider, J., Struble, J., Stentz, A., Taylor, M., Whittaker, W., Wolkowicki, Z., Zhang, W., Ziglar, J.: Autonomous driving in urban environments: boss and the urban challenge. J. Field Robot. (Special Issue on the 2007 DARPA Urban Challenge, Part I) **25**(8), 425–466 (2008)
16. Hult, R., Tabar, R.S.: Path Planning for Highly Automated Vehicles. Mathesis, Gothenburg (2013)
17. Werling, M., Groll, L., Bretthauer, G.: Invariant trajectory tracking with a full-size autonomous road vehicle. IEEE Trans. Robot. **26**(4), 758–765 (2010)
18. Li, X., Sun, Z., Zhu, Q., Liu, D.: A unified approach to local trajectory planning and control for autonomous driving along a reference path. In: 2014 IEEE International Conference on Mechatronics and Automation (ICMA), pp. 1716–1721 (2014)
19. Zhang, S., Deng, W., Zhao, Q., Sun, H., Litkouhi, B.: Dynamic trajectory planning for vehicle autonomous driving. In: 2013 IEEE International Conference on Systems, Man, and Cybernetics (SMC), pp. 4161–4166 (2013). https://doi.org/10.1109/SMC.2013.709
20. Rupp, A., Stolz, M.: Eine erweitere bewertungsfunktion fr umfassende trajektorienplanung auf autobahnen. pp. 337–356. VDI Verlag GMBH, Deutschland (2016)
21. Schramm, D., Hiller, M., Bardini, R.: Single Track Models, pp. 223–253. Springer Berlin Heidelberg, Berlin, Heidelberg (2014)

Chapter 9
Automation of Road Intersections Using Distributed Model Predictive Control

Alexander Katriniok, Peter Kleibaum and Martina Joševski

Abstract The automation of road intersections is increasingly considered as an inevitable next step toward a higher level of autonomy on our roads. For the particular case of fully automated vehicles, we propose a distributed model predictive control approach in which multiple agents are able to pass the intersection simultaneously while keeping a sufficient safety distance to conflicting agents. Therefore, each agent solves a local optimization problem subject to non-convex safety constraints which couple the agents. In order to handle these coupling constraints, we propose constraint prioritization. With that methodology, for two pairwise conflicting agents, the safety constraint is only imposed on the agent with lower priority which does not imply any a priori intersection passing order. Finally, we can solve the distributed optimization problem in parallel without any nested iterations. To solve the local non-convex optimization problems, we apply a semidefinite programming relaxation in combination with randomization to obtain appropriate and feasible solutions. A simulation study finally proves the efficacy of our approach.

9.1 Introduction

Advanced driver assistance systems (ADAS) accommodate the increasing traffic complexity and as such support the driver in his daily driving task. Their development and comprehensive testing is a challenging task for engineers today, especially with

---

A. Katriniok (✉)
Ford Research and Innovation Center (RIC), Süsterfeldstr. 200, 52072 Aachen, Germany
e-mail: de.alexander.katriniok@ieee.org

P. Kleibaum
RWTH Aachen University, Aachen, Germany
e-mail: peter.kleibaum@rwth-aachen.de

M. Joševski
Institute of Automatic Control, RWTH Aachen University,
Steinbachstr. 54, 52074 Aachen, Germany
e-mail: martina.josevski@gmail.com

© Springer International Publishing AG, part of Springer Nature 2019
H. Waschl et al. (eds.), *Control Strategies for Advanced Driver Assistance Systems and Autonomous Driving Functions*, Lecture Notes in Control and Information Sciences 476, https://doi.org/10.1007/978-3-319-91569-2_9

a continuously increasing level of automation. With an increased level of automation, though, there is also significant potential to reduce road congestion, improve efficiency, and increase throughput. In particular, road intersections exhibit remarkable potential for improvement when additionally leveraging vehicle-to-everything (V2X) communication. In the remainder of this chapter, we will highlight the challenges that arise with intersection automation at the design stage. Thinking of getting those systems into series production one day translates among others into the problem of designing algorithmic solutions that can even be executed on an embedded electronic control unit (ECU).

The problem of safely coordinating vehicles in the vicinity of an intersection has extensively been discussed in recent years. A comprehensive overview hereof along with the challenges that arise herewith can be found in [1, 2]. Generally, these approaches can be compared from two different viewpoints: first, according to the controller structure and, second, according to the methodology applied to solve the control problem. From a structural viewpoint, we can distinguish between three main approaches: first, schemes that utilize a central coordination regime that plans and optimizes the trajectory of the vehicles passing through the intersection [3–5]; second, decentralized or distributed schemes that are able to resolve conflicts by negotiating actions without any centralized coordination instance [6–8]; and third, hybrid approaches which combine a centralized regime, e.g., for assigning a passing order to the vehicles, with local decentralized controllers being responsible for determining appropriate control actions [9, 10]. Focusing on the methodology, the intersection conflict resolution problem has been approached by exploiting algorithms originating among others from hybrid system theory [7, 11, 12], multi-agent systems theory relying on resource reservation protocols [5, 13, 14], scheduling-based approaches [15–17], and those based on optimal respectively model predictive control (MPC) [3, 4, 6, 18]. Following a different approach, the authors in [19] transform the conflict resolution problem in T-intersections into a virtual platooning problem utilizing a (virtual) cooperative adaptive cruise control (CACC) system.

We consider MPC to be an appropriate methodology to address such kind of coordination problems that have to deal with constraints explicitly and that have to incorporate anticipated trajectories of conflicting vehicles (also referred to as agents). In the field of MPC-based approaches, [3] proposes a centralized control scheme which is based on collision risk minimization [10, 18] introduce a hierarchical control framework that combines an MPC-based motion planner with a centralized coordination layer that prescribes a certain intersection crossing order to the agents and blocks the intersection exclusively for a passing agent. In a similar way, [4, 9] optimize the intersection passing order for the agents through a central coordination regime, while the local subproblems can be solved locally by the agents. Again, the agents are granted exclusive access to the intersection to pass through. The same idea is adopted by [6], introducing a decision order-based decentralized control concept. Thereby, each agent decides in a sequential order whether to pass the intersection before or after the agents with higher decision order. This sequential approach is based on nested iterations which require information to be exchanged more than once in a single sampling step to solve the decentralized control problem. Another sequential approach is presented in [8] where a priority-based decentralized MPC framework is utilized that controls the agents such that an a priori fixed intersection passing order is maintained. The optimization problem is solved sequentially, propagating the solution of the local optimization problems to agents with lower priority.

In this contribution,[1] we will consider the case of fully automated vehicles using vehicle-to-vehicle (V2V) communication without any roadside unit being involved. Our main objective is to enable the vehicles to pass the intersection simultaneously without blocking it exclusively and thus to increase throughput while keeping an appropriate safety distance to other conflicting vehicles. Therefore, we will propose a distributed MPC-based control approach which allows to solve the distributed optimal control problem (OCP) in a fully parallelized fashion without the need of nested iterations. We will introduce prioritized safety constraints which accommodate the situation that the local subsystems are coupled by the safety constraints. The prioritization regime causes that vehicles with higher priority are granted right-of-way in case of a potential collision without prescribing any intersection passing order, thus being less restrictive. Finally, the resulting distributed non-convex OCP is solved using semidefinite relaxations of the original problem in combination with randomization to obtain feasible and appropriate solutions. The main advantages of our automation scheme compared to previously proposed MPC-based concepts are:

- Multiple vehicles are able to enter and pass the intersection simultaneously instead of blocking it exclusively. In this way, we reduce conservatism and permit a higher traffic throughput.
- The distributed OCP is solved in a fully parallelized fashion without requiring any nested iterations as they are present in a sequential scheme. This property is an inevitable requirement to apply such a distributed scheme in an actual vehicle.
- Constraint prioritization is a generic and straightforward approach without prescribing any intersection passing order. As such, we increase the number of feasible solutions and consequently reduce conservatism.
- In the formulation of the local OCPs, we reduce the number of non-convex constraints to a minimum such that we are able to solve the optimization problem in real time on the given simulation platform.

The remainder of this article is structured as follows: Section 9.2 introduces the kinematic vehicle model, being applied for predictive control, along with the definition of inter-vehicle distances in the vicinity of the intersection. Starting with the centralized conflict resolution problem in Sect. 9.3 while Sect. 9.4 outlines how the centralized problem is decomposed into a distributed problem which is solved locally by the agents. Local feasibility and global feasibility, as well as optimality of the distributed control approach, are then discussed in Sect. 9.5. In Sect. 9.6, we provide a comprehensive simulation study that demonstrates the efficacy of our approach. Finally, we will conclude the article and give an outlook on future work in Sect. 9.7.

---

[1]This article extends the previous publication [20] published by the International Federation of Automatic Control (IFAC).

9.2 Modeling

For controller design, the intersection scenario requires to be represented by a dynamic system model which will be introduced in the remainder of this section. In our approach, we particularly focus on scenarios which consider a single intersection; see Fig. 9.1. Moreover, we have made the following assumptions:

- *Level of Automation*: All vehicles are fully autonomous.
- *Level of Equipment*: All vehicles are equipped with V2V communication.
- *Vehicle Model*: A kinematic vehicle model describing pure longitudinal vehicle motion is utilized.
- *Vehicles per Direction*: Only one vehicle is assumed to approach the intersection from each direction.
- *Crossing Trajectory*: The desired route and desired velocity of every vehicle passing the intersection are a priori known through a high-level long-term planning algorithm which is not the subject of this contribution.
- *Communication Delay*: Data that has been broadcasted via V2V at time k is available to every other vehicle at the next time step $k+1$.
- *Positioning Uncertainties*: Vehicle positions are known precisely and are not subject to uncertainty.

Despite several restrictions, the reader should be aware that these can also be incorporated into our methodology but have intentionally been disregarded to reduce

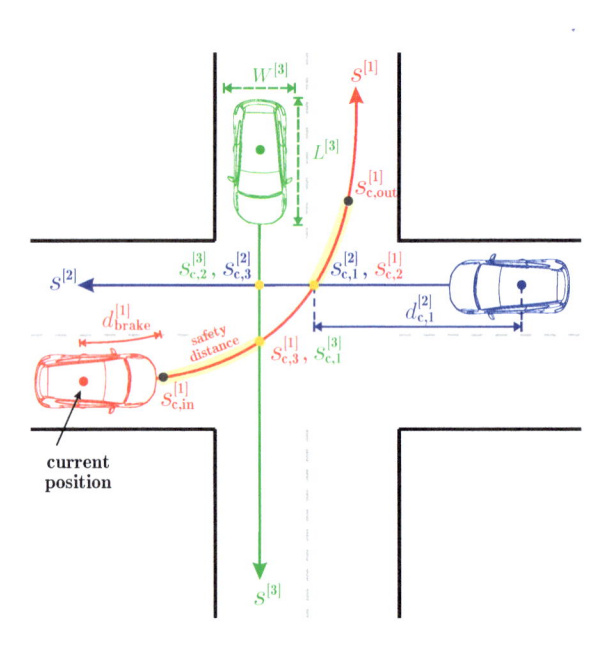

Fig. 9.1 Model of the intersection conflict resolution problem, from [20] © 2017 IFAC, reproduced with permission

complexity. Finally, with the assumption of an a priori known intersection crossing trajectory, we are able to reduce problem dimensionality to a one-dimensional motion problem; i.e., we solely have to consider the movement of a vehicle along its path coordinate.

9.2.1 Vehicle Kinematics

Basically, the intersection scenario consists of a finite set of N_A vehicles or agents $\mathcal{A} \triangleq \{1, \ldots, N_A\}$. By referring to agents, we are adopting the terminology being used in the field of distributed and networked control. The geometry of each agent $i \in \mathcal{A}$ is given by its length $L^{[i]}$ and width $W^{[i]}$. With the simplification to a one-dimensional motion problem, the agent's dynamic behavior can be formulated in a pure kinematic fashion in terms of its acceleration $a_x^{[i]}$, velocity $v^{[i]}$, and path coordinate $s^{[i]}$. Here, the path coordinate refers to the position of the vehicle's geometric center in the agent's local reference frame; see Fig. 9.1. Particularly, we apply a double-integrator model to obtain the agent's velocity by integrating its acceleration and finally the path coordinate by integrating the velocity along the path. To reflect the ability of each agent to accelerate and decelerate, we summarize drivetrain dynamics as a first-order lag element with dynamic time constant $T_{a_x}^{[i]}$. Thus, we can state the kinematic model of every agent in terms of the linear time-invariant model

$$\Sigma^{[i]} \triangleq \begin{cases} \dot{x}^{[i]} = A^{[i]} x^{[i]} + B^{[i]} u^{[i]} \\ y^{[i]} = C^{[i]} x^{[i]} \end{cases} \quad (9.1)$$

with $A^{[i]} = \begin{bmatrix} -1/T_{a_x}^{[i]} & 0 & 0 \\ 1 & 0 & 0 \\ 0 & 1 & 0 \end{bmatrix}$, $B^{[i]} = \begin{bmatrix} 1/T_{a_x}^{[i]} \\ 0 \\ 0 \end{bmatrix}$ and $C^{[i]} = \begin{bmatrix} 0 & 1 & 0 \end{bmatrix}$. The state vector $x^{[i]} = [a_x^{[i]}, v^{[i]}, s^{[i]}]^T$ of agent i is composed of its acceleration, velocity, and path coordinate, while the only controlled output is the agent's velocity. The origin of the agent's path coordinate frame $s^{[i]}$ refers to the first collision point with another agent or, if there is none, to the current position when the scenario is initialized. To influence the dynamic behavior of each agent, we introduce the demanded acceleration $u^{[i]} = a_{x,\text{ref}}^{[i]}$ as manipulated variable. Furthermore, states and inputs are assumed to be constrained by polyhedral sets, i.e., $x^{[i]} \in \mathcal{X}^{[i]}$ and $u^{[i]} \in \mathcal{U}^{[i]}$. To apply the prediction model in a control scheme which is based on a finite-dimensional optimization problem, we discretize (9.1) by using a zero-order hold discretization technique such that we gain the following discrete-time linear time-invariant prediction model

$$\Sigma_d^{[i]} \triangleq \begin{cases} x_{k+1}^{[i]} = A_k^{[i]} x_k^{[i]} + B_k^{[i]} u_k^{[i]} \\ y_k^{[i]} = C_k^{[i]} x_k^{[i]} \end{cases} \quad (9.2)$$

9.2.2 Inter-Vehicle Distances

Besides the necessity to predict the evolution of the agents' states, a definition of inter-vehicle distances between two agents $i, l \in \mathcal{A}$ is required. As a prerequisite, we define the collision points $s_{c,i}^{[l]}$ and $s_{c,l}^{[i]}$ as the intersection of spacial trajectories of agents $i, l \in \mathcal{A}$ along the corresponding path coordinates $s^{[i]}$ and $s^{[l]}$; see Fig. 9.1. If these do not exist, we define $s_{c,l}^{[i]} = s_{c,i}^{[l]} = \infty$. As such, the distance of agent i and agent l to their respective collision points $s_{c,l}^{[i]}$ and $s_{c,i}^{[l]}$ can be stated as

$$d_{c,l}^{[i]} = |s^{[i]} - s_{c,l}^{[i]}|, \quad d_{c,i}^{[l]} = |s^{[l]} - s_{c,i}^{[l]}| \tag{9.3}$$

Then, the distance $d_l^{[i]}$ of agent $i \in \mathcal{A}$ to agent $l \in \mathcal{A}$ can be expressed as

$$d_l^{[i]} = \begin{cases} |s^{[i]} - s_{c,l}^{[i]}| + |s^{[l]} - s_{c,i}^{[l]}| & , s_{c,l}^{[i]}, s_{c,i}^{[l]} \neq \infty \\ \infty & , \text{otherwise} \end{cases} \tag{9.4}$$

9.3 Centralized Conflict Resolution Problem

Having established a suitable prediction model, we will subsequently outline how the intersection conflict resolution problem can generally be formulated in terms of a centralized OCP with non-convex safety constraints. Then, the next section focuses on how to decompose the centralized OCP in local OCPs and finally how to solve the non-convex optimization problem in a distributed fashion. For solving the intersection conflict resolution problem, we will introduce an MPC-based scheme. The basic idea of MPC is that at every time step k a finite-time optimal control problem is solved over a finite prediction horizon of length H_p, while controls are applied over a control horizon of length $H_u \leq H_p$. After optimization, only the first control input is applied to the plant and optimization is repeatedly executed over a shifted horizon at the next time step $k + 1$. To deal with predictions in terms of mathematical notation, we will subsequently use $\{\cdot\}_{(k+j|k)}$ to refer to the predicted value of variable $\{\cdot\}$ at the future time step $k + j$ when the current time step is k.

With these preliminaries, we are now able to state the objectives of our control scheme and finally to formalize them in terms of a receding horizon optimization problem. Particularly, the intersection conflict resolution problem has six main objectives. Apparently, the most relevant objective is to avoid any collisions between conflicting agents, i.e., agents with a joint collision point. As such, for agent $i \in \mathcal{A}$ the corresponding set of conflicting agents can be formalized as

$$\mathcal{A}_c^{[i]} = \left\{ l \in \mathcal{A} \mid l \neq i \wedge s_{c,l}^{[i]} \neq \infty \right\}. \tag{9.5}$$

To avoid collisions, it has to be ensured that every two conflicting agents $i, l \in \mathcal{A}$ pass their respective collision point with a minimum safety distance to each other which can finally be stated as a safety constraint for agent i over the entire prediction horizon, i.e.,

$$d^{[i]}_{l,(k+j|k)} \geq d^{[i]}_{safe,l,(k+j|k)}, \quad \forall l \in \mathcal{A}^{[i]}_c, \quad \forall j \in \{1, \ldots, H_p\} \tag{9.6}$$

where $d^{[i]}_{safe,l,(k+j|k)}$ denotes the desired safety distance of agent i to agent l at the future time $k+j$. Here, we have introduced the safety distance as a time-dependent variable to reduce conservatism for conflict situations with turning vehicles. For further details, see Remark 9.3.1. Recalling definition (9.4) of inter-vehicle distances, we have to deal with an absolute value constraint. From the perspective of agent i, agent l should as such either be located in front or behind the ego-vehicle having a minimum safety distance of $d^{[i]}_{safe,l}$. Apparently, such constraint cuts off an inner region of the feasible set which translates into a non-convex constraint that might significantly increase the computational complexity of the OCP. As a second objective, the agent's travel speed $v^{[i]}$ should be as close as possible to the desired speed $v^{[i]}_{ref}$. According to our assumptions in Sect. 9.2, the desired velocity trajectory originates from a high-level long-term planning algorithm. Third, intersection passing should happen as efficient as possible which relates to minimizing fuel consumption. Therefore, we aim at minimizing absolute accelerations. Fourth, the driver should feel comfortable which requires smooth input trajectories. The second to fourth requirement can be stated in terms of a local quadratic objective function for every agent $i \in \mathcal{A}$ as

$$J^{[i]}(x_0^{[i]}, u_{(\cdot|k)}^{[i]}) = Q^{[i]} \sum_{j=1}^{H_p} (v^{[i]}_{ref,(k+j|k)} - v^{[i]}_{(k+j|k)})^2 \tag{9.7}$$

$$+ R^{[i]} \sum_{j=0}^{H_u-1} \Delta u^{[i],2}_{(k+j|k)} + S^{[i]} \sum_{j=0}^{H_u-1} u^{[i],2}_{(k+j|k)}$$

where $x_0^{[i]} = x_k^{[i]}$ denotes the current state at time k, $u^{[i]}_{(\cdot|k)} = [u^{[i]}_{(k|k)}, \ldots, u^{[i]}_{(k+H_u-1|k)}]^T$ the control inputs of agent i over the control horizon, $\Delta u^{[i]}_{(k+j|k)} = u^{[i]}_{(k+j|k)} - u^{[i]}_{(k+j-1|k)}$ the step change of the control input over the control horizon, while $Q^{[i]} > 0, R^{[i]} > 0$ and $S^{[i]} > 0$ are positive weighting coefficients. The last two requirements we have to consider in controller design are related to state and input constraints. Particularly, the fifth requirement accounts for actuator limitations in terms of maximum and minimum feasible accelerations which translates into the input constraint

$$u^{[i]}_{(k+j|k)} \in \mathcal{U}^{[i]} \triangleq \left\{ u \in \mathbb{R} \mid \underline{u}^{[i]} \leq u \leq \overline{u}^{[i]} \right\}, \quad \forall j \in \{0, \ldots, H_u - 1\}. \tag{9.8}$$

Finally, each agent has to obey the legal speed limits such that its maximum velocity has to be constrained. Furthermore, agents should only drive in the forward direction.

Therefore, we impose a lower velocity bound of zero. These restrictions can then be summarized as a state constraint of the form

$$x_{(k+j|k)}^{[i]} \in \mathcal{X}_{(k+j|k)}^{[i]} \triangleq \left\{ \xi \in \mathbb{R}^3 \mid 0 \leq [\xi]_2 \leq \bar{v}_{(k+j|k)}^{[i]} \right\}, \; \forall j \in \{1, \ldots, H_p\}. \quad (9.9)$$

where $[\xi]_2$ refers to the velocity as second system state. For feasibility reasons, we finally have to ensure that the prediction horizon at least covers the coordinate set $\mathcal{S}_c^{[i]} \triangleq [s_{c,in}^{[i]}, s_{c,out}^{[i]}]$ in which potential collisions might occur between agents during intersection crossing. If the scenario is not feasible, we want the agent to conduct an emergency braking maneuver in a brake safe distance $d_{brake}^{[i]}$, i.e., when entering the set $\mathcal{S}_{cb}^{[i]} \triangleq [s_{c,in}^{[i]} - d_{brake}^{[i]}, s_{c,out}^{[i]}]$; see Fig. 9.1. Therefore, the minimum mean velocity of agent $i \in \mathcal{A}$ is bounded by

$$\frac{1}{H_p+1} \sum_{j=0}^{H_p} v_{(k+j|k)}^{[i]} \geq \underline{v}_{mean}^{[i]} \quad (9.10)$$

to ensure that the prediction horizon at least covers $\mathcal{S}_{cb}^{[i]}$. If $s_k^{[i]} \in \mathcal{S}_{cb}^{[i]}$, we determine $\underline{v}_{mean}^{[i]}$ by dividing the remaining distance to $s_{c,out}^{[i]}$ by the preview time covered by the prediction horizon. Otherwise, it is set to zero. Finally, with these requirements we are able to summarize the centralized OCP as

Centralized Optimal Control Problem

$$\underset{\bar{u}_{(\cdot|k)}}{\text{minimize}} \quad \sum_{i \in \mathcal{A}} J^{[i]}(x_0^{[i]}, u_{(\cdot|k)}^{[i]}) \quad (9.11)$$

subject to   system dynamics (9.2)

safety constraints (9.6)

state constraints (9.9)–(9.10)

input constraints (9.8)

with $\bar{u}_{(\cdot|k)} = [u_{(\cdot|k)}^{[1],T}, \ldots, u_{(\cdot|k)}^{[N_A],T}]^T$. As already stated, safety constraint (9.6) renders the OCP into a non-convex one which might be computationally demanding when it should be solved to optimality. In the next section, we will therefore come up with a relaxation of the non-convex safety constraints in order to solve the distributed OCP in a computationally tractable way.

Remark 9.3.1 In (9.6), we have introduced the safety distance $d_{safe,l,(k+j|k)}^{[i]}$ as a time-dependent quantity. If both conflicting agents are crossing the intersection straight, a constant safety distance is applied. However, if one of the agents is turning, the required safety distance is larger when both agents approach the joint collision point and smaller when one of the agents has just passed it. This observation results from the rotational movement of the agents. Consequently, to reduce conservatism in case of turning agents, we have introduced the safety distance as a time-varying parameter that depends on the position of both agents.

9.4 Distributed Conflict Resolution Problem

The previous section has outlined how the coordination problem can generally be formulated in terms of a centralized OCP. This section, though, will pinpoint how the large centralized problem can be distributed by splitting it into smaller pieces such that each agent solves its own local OCP.

9.4.1 Primal Decomposition

To solve the conflict resolution problem in a distributed fashion, we decompose the centralized OCP in local OCPs by applying a primal decomposition technique. Apparently, the cost function of the centralized OCP (9.11) can easily be decomposed such that each agent $i \in \mathcal{A}$ minimizes the local objective function $J^{[i]}(x_0^{[i]}, u_{(\cdot|k)}^{[i]})$. In a similar fashion, we can independently account for system dynamics, input, and state constraints. However, the safety constraints are imposed pairwise on each of two conflicting agents $i, l \in \mathcal{A}$ and as such couple these local subsystems. The main issues that arise in the formulation of the distributed OCP are related to these joint constraints. Particularly, we have to deal with the following issues:

1. According to (9.6), safety constraints are absolute value constraints. These have to be reformulated appropriately to obtain a suitable form which can be handled by a numerical solver. At the same time, this reformulation has to allow for an efficient solution of the optimization problem.
2. A simple decomposition results in safety constraints that are imposed pairwise on both conflicting agents. However, in a distributed scheme with non-convex coupled constraints this approach might cause convergence issues as discussed in Sect. 9.4.2.

In the remainder of this section, we will propose a solution to both of these issues step by step and finally come up with a formulation of a distributed OCP which can be solved in a fully parallelized fashion.

First, we are going to outline how to reformulate safety constraints such that these are suitable for a numerical solver and how to minimize the required computational burden that arises with these non-convex constraints. According to the definition of inter-vehicle distances (9.4), safety constraint (9.6) for agent $i \in \mathcal{A}$ can be rewritten as

$$|s_{(k+j|k)}^{[i]} - s_{c,l}^{[i]}| \geq d_{safe,l,(k+j|k)}^{[i]} - d_{c,i,(k+j|k)}^{[l]}, \quad \forall l \in \mathcal{A}_c^{[i]}, \; \forall j \in \{1, \ldots, H_p\} \quad (9.12)$$

where $|s^{[i]}_{(k+j|k)} - s^{[i]}_{c,l}|$ and $d^{[l]}_{c,i,(k+j|k)} = |s^{[l]}_{(k+j|k)} - s^{[l]}_{c,i}|$ denote the distances of agent $i \in \mathcal{A}$ and $l \in \mathcal{A}^{[i]}_c$ to their respective collision points at time $k+j$ over the prediction horizon. To utilize the distance of agent l to its respective collision point in the local OCP of agent i, we assume $d^{[l]}_{c,i,(k+j|k)}$ to be broadcasted by agent l using V2V communication. According to [21], there are basically two approaches to reformulate safety constraint (9.12) such that it can be handled by a numerical solver:

1. Replace every absolute value constraint by two separate linear constraints such that optimization has to choose one of them. In this way, a mixed-integer OCP is obtained.
2. Transform every absolute value constraint into a quadratic one.

While methodology (1) is frequently applied in the literature, we exploit method (2) to solve the OCP even more efficiently. With methodology (1), we have to choose between two linear constraints for each safety constraint such that additional binary selection variables have to be incorporated into the optimization problem. While [21] also requires additional optimization variables utilizing methodology (2), our approach allows to get along without increasing the dimensionality of the optimization problem. When analyzing the absolute value safety constraint (9.12), we can recognize that it is satisfied in all cases if the distance of agent $l \in \mathcal{A}^{[i]}_c$ to its collision point is larger than or equal to the safety distance, i.e., $d^{[l]}_{c,i,(k+j|k)} \geq d^{[i]}_{safe,l,(k+j|k)}$. As such, we only have to impose the safety constraint if $d^{[l]}_{c,i,(k+j|k)} < d^{[i]}_{safe,l,(k+j|k)}$ which significantly reduces the required number of non-convex constraints over the prediction horizon. As both sides of the inequality constraint (9.12) are then larger or equal to zero, we can square both sides of the safety constraint, thus gaining a non-convex quadratic inequality constraint of the form

$$(s^{[i]}_{(k+j|k)} - s^{[i]}_{c,l})^2 \geq (d^{[i]}_{safe,l,(k+j|k)} - d^{[l]}_{c,i,(k+j|k)})^2, \tag{9.13}$$
$$\forall l \in \mathcal{A}^{[i]}_c : d^{[l]}_{c,i,(k+j|k)} < d^{[i]}_{safe,l,(k+j|k)}, j \in \{1, \ldots, H_p\}$$

which now has not to be imposed for every time step of the prediction horizon. This formulation will contribute to obtain local OCPs which can be solved more efficiently.

When imposing these safety constraints in a discrete-time MPC formulation, we can assure that no collisions with other agents occur at any discrete time step. In between two time steps, though, that is not necessarily the case, especially for larger sampling times. When illustrating the path coordinate $s^{[i]}$ of agent i over time, the required safety distance to a conflicting agent l can be represented as a safety polygon around the path coordinate trajectory of agent l that must not be intersected by the trajectory of agent i. A safety constraint violation might occur if an *edge* of the safety polygon is intersected between two subsequent time steps $k+j$ and $k+j+1$. Therefore, the required safety distance $d^{[i]}_{safe,l,(k+j|k)}$ is adapted accordingly in those cases to ensure continuous-time safety constraint satisfaction.

Summarizing, we can render the distributed OCP for every agent $i \in \mathcal{A}$ as a non-convex quadratically constrained quadratic program (QCQP) subject to pairwise safety constraints:

Distributed Optimal Control Problem (w/ pairwise safety constraints), $\forall i \in \mathcal{A}$:

$$\underset{u^{[i]}_{(\cdot|k)}}{\text{minimize}} \quad J^{[i]}(x_0^{[i]}, u^{[i]}_{(\cdot|k)}) \tag{9.14}$$

subject to system dynamics (9.2)

safety constraints (9.13)

state constraints (9.9)–(9.10)

input constraints (9.8)

Remark 9.4.1 Without loss of generality, we assume for every two conflicting agents $i, l \in \mathcal{A}$ that $d^{[i]}_{safe,l,(k+j|k)} = d^{[l]}_{safe,i,(k+j|k)}$ for $j \in \{1, \ldots, H_p\}$. If this is not the case, we have to take the maximum of both safety distances to obtain the same optimization problem after decomposition as in the centralized case.

Remark 9.4.2 We assume $d^{[l]}_{c,i,(\cdot|k)} = [d^{[l]}_{c,i,(k+1|k)}, \ldots, d^{[l]}_{c,i,(k+H_p|k)}]^T$ to be broadcasted by every agent l at every time step k through V2V communication. Moreover, it is assumed that $d^{[l]}_{c,i,(\cdot|k)}$ is available to agent i at time k when it has been broadcasted by agent l at time $k-1$. The trajectory that is actually received by agent i is: $d^{[l]}_{c,i,(k|k-1)}, \ldots, d^{[l]}_{c,i,(k+H_p-1|k-1)}$. For optimization purposes, however, this trajectory is required from time step $k+1$ to $k+H_p$. Therefore, agent l extrapolates its trajectory for time step $k+H_p+1$ at time $k-1$ using the kinematic prediction model (9.2) assuming that the control input at end of the control horizon remains constant for that prediction step. Finally, at time k agent i is able to exploit the entire predicted trajectory $d^{[l]}_{c,i,(k|k-1)}, \ldots, d^{[l]}_{c,i,(k+H_p|k-1)}$ of agent l for optimization purposes.

9.4.2 Prioritizing Coupled Safety Constraints

Solving the distributed OCP (9.14) with coupled constraints is a challenging task due to its non-convex nature. For convex problems with separable objective functions and coupled constraints, there are various mature methods like distributed subgradient methods [22, 23], proximal methods [24], or the alternating direction method of multipliers [25] to solve the convex distributed problem. In these approaches, the Lagrangian multiplier of the coupled constraints is basically handled as a consensus variable during optimization such that consensus is obtained on satisfying constraints which couple two or more local subsystems. From an optimization perspective, these methods can additionally be subdivided into variants that are solved in a sequential fashion, thus requiring nested iterations, and those that allow for fully parallelized computations. When aiming at experimental evaluations, nested iterations will cause additional communication effort and might thus prohibitively increase the overall computation time for solving the optimization problem between two subsequent time steps. Consequently, for the intersection conflict resolution problem fully parallelized computations are preferable. While these mature methods are available for convex distributed problems, for non-convex distributed problems, however, there is no unique way

1. how to handle coupling constraints and
2. how to deal with the non-convexity of the optimization problem.

The remainder of this section will focus on the first subject, while the second will be addressed in Sect. 9.4.4. In the decomposed OCP (9.14), we have imposed safety constraints pairwise on conflicting agents. Without any consensus-based approach for a non-convex distributed OCP (to the best knowledge of the authors, there is no unique and mature method available), imposing safety constraints pairwise might lead to deadlock-like situations or even worse to collisions when solving the problem in a fully parallelized fashion. But what is the reason for such behavior? Apparently, when there is no common consensus among agents, the intention of each agent might change over time, just like two people approaching each other on the sidewalk. From an MPC perspective, control and state trajectories of an agent do not necessarily converge such that the other agent can adapt and finally global convergence is reached. To cope with this issue while being able to solve the non-convex distributed OCP in parallel without any nested iterations, we propose to adapt prioritized safety constraints instead of pairwise constraints. The basic idea is to introduce priorities on agents such that just one of two conflicting agents has to be aware of the joint safety constraints while these joint constraints are dropped for the other agent. In this way, we force a consensus between agents. This idea is summarized in the following definition.

Definition 9.4.1 (*Prioritized Safety Constraints*) Define the bijective prioritization function $\gamma : \mathcal{A} \to \mathbb{N}_+$ which assigns a unique passing priority to every agent $i \in \mathcal{A}$ where a lower value corresponds to a higher priority. Particularly, agent $l \in \mathcal{A}_c^{[i]}$ is permitted to pass its corresponding collision point $s_{c,i}^{[l]}$ without considering agent $i \in \mathcal{A}$ if and only if agent l has a higher priority than agent i, i.e., $\gamma(l) < \gamma(i)$. As such, we can drop safety constraint (9.13) for agent l, while only agent i has to be aware of it. By defining the prioritized conflict set

$$\mathcal{A}_{c,\gamma}^{[i]} = \left\{ l \in \mathcal{A}_c^{[i]} \mid l \neq i \wedge \gamma(l) < \gamma(i) \wedge s_{c,l}^{[i]} \neq \infty \right\} \tag{9.15}$$

which contains the agents l having a higher priority than agent i and a joint collision point with agent i, we can finally rewrite (9.13) for agent i to

$$(s_{(k+j|k)}^{[i]} - s_{c,l}^{[i]})^2 \geq (d_{safe,l,(k+j|k)}^{[i]} - d_{c,i,(k+j|k)}^{[l]})^2, \tag{9.16}$$

$$\forall l \in \mathcal{A}_{c,\gamma}^{[i]} : d_{c,i,(k+j|k)}^{[l]} < d_{safe,l,(k+j|k)}^{[i]}, \ j \in \{1, \ldots, H_p\}.$$

The general idea of prioritizing agents has also been leveraged in [6, 8, 10]. However, in these approaches prioritization directly implies an intersection passing order. In our approach, prioritizing agents and finally safety constraints does not prescribe any intersection passing order which would exclude potentially feasible solutions a priori. Instead, prioritized constraints correspond to a virtual traffic rule that claims that an agent with lower priority has to give right-of-way to the agent with higher priority in case of a potential conflict. Moreover, our approach still allows the low-priority agent to pass the intersection before the high-priority agent. When substituting safety constraints (9.13) in the distributed OCP (9.14) with their prioritized counterparts (9.16), we finally gain the

Prioritized Distributed Optimal Control Problem, $\forall i \in \mathcal{A}$:

$$\underset{u^{[i]}_{(\cdot|k)}}{\text{minimize}} \quad J^{[i]}(x^{[i]}_0, u^{[i]}_{(\cdot|k)}) \tag{9.17}$$

subject to system dynamics (9.2)

prioritized safety constraints (9.16)

state constraints (9.9)–(9.10)

input constraints (9.8)

The resulting distributed OCP can now be solved in a fully parallelized fashion. Recapitulating, until now we have decomposed the centralized OCP, have reformulated safety constraints, and have come up with a solution for the issue of coupled constraints in the distributed setting. Now, we still have to solve the local non-convex OCPs for every agent $i \in \mathcal{A}$ which will be addressed in the next section.

9.4.3 Prioritization Functions

The prioritization approach, introduced in Sect. 9.4.2, is generic such that prioritization functions $\gamma : \mathcal{A} \to \mathbb{N}_+$ can be chosen arbitrarily, e.g., according to traffic rules or an optimization criteria in a higher-level optimization problem. Reference [26] proposes various criteria like first-in first-out (FIFO), the distance to the intersection and the time to react. From the authors' point of view, the time to react is a very appropriate criteria for the conflict resolution problem as it accounts for the ability of every agent to stop before entering a conflict region in the intersection. In our case, we adapt this idea such that the time to react $t^{[i]}_{TTR}$ of agent i refers to the time that is left until the agent is not able to stop before entering the conflict region, i.e., the coordinate set $\mathcal{S}^{[i]}_c$. With less reaction time should result in a higher priority, we can deduce the particular prioritization function which is utilized in the remainder of the contribution as

$$t^{[i]}_{TTR} < t^{[l]}_{TTR} \Rightarrow \gamma(i) < \gamma(l), \quad \forall i, l \in \mathcal{A}, \; i \neq l. \tag{9.18}$$

To be independent of any central priority coordination regime, we assume that priorities are negotiated once when the scenario is established at time t_0 and then remain time-invariant for the entire intersection crossing maneuver. In our case, the vehicles would initially exchange their paths through the intersection and propagate their time to react with respect to $\mathcal{S}_c^{[i]}$. With the assumption of time-invariant priorities, we are able to ensure convergence of the distributed OCP as the agents with lower priority are able to adapt to the agents with higher priority.

9.4.4 Solving the Distributed Non-convex OCP

To approach a solution of the local non-convex optimization problems, we have to apply several transformations to our distributed OCP (9.17) with prioritized constraints. Recalling, the local objective function (9.7) of every agent $i \in \mathcal{A}$ and its safety constraints (9.16) are quadratic functions. Generally, a quadratic function depending on variable $u \in \mathbb{R}^n$, $n \in \mathbb{N}_+$ can be represented by $u^T P u + q^T u + r$ for some matrix P and vector q of appropriate dimension, while r is a scalar value. If P is positive definite, i.e., for $P \succeq 0$, the quadratic function is said to be convex, for a negative definite matrix $P \preceq 0$ or an indefinite P, it is referred to as a non-convex function. Thus, we can rewrite (9.17) for every agent $i \in \mathcal{A}$ as a

Quadratically Constrained Quadratic Program, $\forall i \in \mathcal{A}$:

$$\underset{u_{(\cdot|k)}^{[i]}}{\text{minimize}} \quad u_{(\cdot|k)}^{[i],T} P_0^{[i]} u_{(\cdot|k)}^{[i]} + q_0^{[i],T} u_{(\cdot|k)}^{[i]} + r_0^{[i]} \tag{9.19}$$

$$\text{subject to} \quad u_{(\cdot|k)}^{[i],T} P_{l,(k+j|k)}^{[i]} u_{(\cdot|k)}^{[i]} + q_{l,(k+j|k)}^{[i],T} u_{(\cdot|k)}^{[i]} + r_{l,(k+j|k)}^{[i]} \leq 0,$$

$$\forall l \in \mathcal{A}_{c,\gamma}^{[i]} : d_{c,i,(k+j|k)}^{[l]} < d_{safe,l,(k+j|k)}^{[i]}, j \in \{1, \ldots, H_p\}$$

state constraints (9.9)–(9.10)

input constraints (9.8)

where $P_0^{[i]} \succeq 0$, $P_{l,(k+j|k)}^{[i]} \preceq 0$, $q_0^{[i]}$, and $q_{l,(k+j|k)}^{[i]}$ are matrices and vectors of appropriate dimensions, while $r_0^{[i]}$ and $r_{l,(k+j|k)}^{[i]}$ are scalar values. The first constraint represents the reformulated safety constraint of agent i being in conflict with agent l at time $k+j$. In this context, the negative semidefinite matrix $P_{l,(k+j|k)}^{[i]}$ reflects the non-convexity of safety constraints. As a next step, we define $U^{[i]} \triangleq u_{(\cdot|k)}^{[i]} u_{(\cdot|k)}^{[i],T}$ in accordance to [27] and recast problem (9.19) into a

Rank-Constrained Semidefinite Program, $\forall i \in \mathcal{A}$:

$$\underset{U^{[i]}, u^{[i]}_{(\cdot|k)}}{\text{minimize}} \quad \text{Tr}(U^{[i]} P_0^{[i]}) + q_0^{[i],T} u^{[i]}_{(\cdot|k)} + r_0^{[i]} \tag{9.20}$$

$$\text{subject to} \quad \text{Tr}(U^{[i]} P_{l,(k+j|k)}^{[i]}) + q_{l,(k+j|k)}^{[i],T} u^{[i]}_{(\cdot|k)} + r_{l,(k+j|k)}^{[i]} \leq 0,$$

$$\forall l \in \mathcal{A}_{c,\gamma}^{[i]} : d_{c,i,(k+j|k)}^{[l]} < d_{safe,l,(k+j|k)}^{[i]}, \, j \in \{1, \ldots, H_p\}$$

$$\mathcal{S}^{[i]} \triangleq \begin{bmatrix} U^{[i]} & u^{[i]}_{(\cdot|k)} \\ u^{[i],T}_{(\cdot|k)} & 1 \end{bmatrix} \succeq 0, \, \text{rank}(\mathcal{S}^{[i]}) = 1$$

state constraints (9.9)–(9.10)

input constraints (9.8)

where $\text{Tr}(M)$ stands for the trace of a square matrix M and $\mathcal{S}^{[i]}$ denotes the Schur complement [27] of $U^{[i]} - u^{[i]}_{(\cdot|k)} u^{[i],T}_{(\cdot|k)}$. Through constraint $\mathcal{S}^{[i]} \succeq 0$, $\text{rank}(\mathcal{S}^{[i]}) = 1$, we claim that $U^{[i]} = u^{[i]}_{(\cdot|k)} u^{[i],T}_{(\cdot|k)}$ is satisfied. As such, in this problem representation we still have the same problem as in (9.17) and (9.19). However, the resulting rank-constrained problem might be even harder to solve. By finally dropping the rank-1 constraint, we obtain a semidefinite programming relaxation (SDR) of the original problem which can now be solved efficiently with mature optimization techniques:

Semidefinite Programming Relaxation of (9.20), $\forall i \in \mathcal{A}$:

$$\underset{U^{[i]}, u^{[i]}_{(\cdot|k)}}{\text{minimize}} \quad \text{Tr}(U^{[i]} P_0^{[i]}) + q_0^{[i],T} u^{[i]}_{(\cdot|k)} + r_0^{[i]} \tag{9.21}$$

$$\text{subject to} \quad \text{Tr}(U^{[i]} P_{l,(k+j|k)}^{[i]}) + q_{l,(k+j|k)}^{[i],T} u^{[i]}_{(\cdot|k)} + r_{l,(k+j|k)}^{[i]} \leq 0,$$

$$\forall l \in \mathcal{A}_{c,\gamma}^{[i]} : d_{c,i,(k+j|k)}^{[l]} < d_{safe,l,(k+j|k)}^{[i]}, \, j \in \{1, \ldots, H_p\}$$

$$\mathcal{S}^{[i]} \triangleq \begin{bmatrix} U^{[i]} & u^{[i]}_{(\cdot|k)} \\ u^{[i],T}_{(\cdot|k)} & 1 \end{bmatrix} \succeq 0$$

state constraints (9.9)–(9.10)

input constraints (9.8)

Although we are able to gain the optimal solution of the convex optimization problem (9.21), SDR just provides a lower bound on the optimal cost of the original non-convex OCP (9.19) and not necessarily a good or even feasible solution of (9.19), see [27]. As such, there are two possible cases after solving the SDR (9.21):

1. We have solved the original non-convex OCP (9.19) to optimality. Then, for the rank of Schur complement matrix $\mathcal{S}^{[i]}$ holds: $\text{rank}(\mathcal{S}^{[i]}) = 1$.
2. We have not solved the original problem and obtain $\text{rank}(\mathcal{S}^{[i]}) > 1$. Then, we have to prove if the solution is feasible and to investigate whether we are able to find better solutions.

To actually assess the rank of $\mathcal{S}^{[i]}$, a pure rank test might cause numerical issues. Therefore, we analyze the singular values of $\mathcal{S}^{[i]}$ instead to decide the rank of the matrix. Let $\sigma_j(\mathcal{S}^{[i]})$ denote the jth largest singular value of $\mathcal{S}^{[i]}$. Then, we consider $\mathcal{S}^{[i]}$ to have rank 1 if the second largest singular value is less than a threshold, i.e., $\sigma_2(\mathcal{S}^{[i]}) \leq \sigma_{2,thld}^{[i]}$.

Finally, we have to discuss how the situation of rank$(\mathcal{S}^{[i]}) > 1$ can be accommodated, i.e., when a suboptimal or even an infeasible solution with respect to the original problem has been gained. According to the literature, this issue can be addressed using rank minimization strategies [28], convex restriction [29], or randomization techniques [29]. While rank minimization strategies might be computationally demanding and convex restrictions increase conservatism by excluding feasible solutions, we apply a variant of the randomization technique described in [29] to yield an appropriate and feasible solution of the original problem (9.19). Assuming $(U^{[i]*}, u_{(\cdot|k)}^{[i]*})$ to be the optimal solution of the relaxed problem (9.21), we can interpret $u_{(\cdot|k)}^{[i]*}$ as a mean value and $U^{[i]*} - u_{(\cdot|k)}^{[i]*} u_{(\cdot|k)}^{[i]*,T}$ as a covariance matrix when applying randomization techniques. Then, we can draw $N_{samples}$ samples $\tilde{u}_{(\cdot|k)}^{[i]}$ by considering $\tilde{u}_{(\cdot|k)}^{[i]}$ as a random variable being Gaussian distributed, i.e., $\tilde{u}_{(\cdot|k)}^{[i]} \sim \mathcal{N}(u_{(\cdot|k)}^{[i]*}, U^{[i]*} - u_{(\cdot|k)}^{[i]*} u_{(\cdot|k)}^{[i]*,T})$. Finally, we can solve the non-convex QCQP (9.19) on average over this distribution, thus yielding the

Randomized Non-Convex QCQP, $\forall i \in \mathcal{A}$:

$$\underset{\tilde{u}_{(\cdot|k)}^{[i]}}{\text{minimize}} \quad \mathbb{E}[\tilde{u}_{(\cdot|k)}^{[i],T} P_0^{[i]} \tilde{u}_{(\cdot|k)}^{[i]} + q_0^{[i],T} \tilde{u}_{(\cdot|k)}^{[i]} + r_0^{[i]}] \tag{9.22}$$

$$\text{subject to} \quad \mathbb{E}[\tilde{u}_{(\cdot|k)}^{[i],T} P_{l,(k+j|k)}^{[i]} \tilde{u}_{(\cdot|k)}^{[i]} + q_{l,(k+j|k)}^{[i],T} \tilde{u}_{(\cdot|k)}^{[i]} + r_{l,(k+j|k)}^{[i]}] \leq 0,$$

$$\forall l \in \mathcal{A}_{c,\gamma}^{[i]} : d_{c,i,(k+j|k)}^{[l]} < d_{safe,l,(k+j|k)}^{[i]}, j \in \{1, \ldots, H_p\}$$

state constraints (9.9)–(9.10)

input constraints (9.8)

During randomization, we prove constraint satisfaction for every sample $\tilde{u}_{(\cdot|k)}^{[i]}$ and finally choose the sample with the lowest value of the cost function. However, in some scenarios the feasible set might be very tight which obviously complicates finding a solution using randomization techniques unless a very large number of samples is used. To ensure feasibility, we propose to evaluate safety and velocity constraints as soft constraints. For the velocity constraints, we particularly define the lower velocity bound (i.e., standstill) as hard constraint, while the upper bound is formulated as a soft constraint. When implementing the lower bound as soft constraint, the agents occasionally tend to drive backward to avoid safety constraint violations which is actually not desired. In case of constraint violation during randomization, each violation of constraints contributes to the cost function. Therefore, we define the violation cost

$$J_{viol}^{[i]}(x_0^{[i]}, \tilde{u}_{(\cdot|k)}^{[i]}) = \rho_d^{[i]} \sum_{(j,l) \in \mathcal{C}_d^{[i]}} \varepsilon_{d,l,(k+j|k)}^{[i]} + \rho_v^{[i]} \sum_{j \in \mathcal{C}_v^{[i]}} \varepsilon_{v,(k+j|k)}^{[i]} \tag{9.23}$$

which summarizes the weighted constraint violations over the prediction horizon where $\rho_d^{[i]} > 0$ and $\rho_v^{[i]} > 0$ are appropriately chosen positive weighting factors. Here, $\mathcal{C}_d^{[i]}$ and $\mathcal{C}_v^{[i]}$ refer to all time steps $j \in \{1, \ldots, H_p\}$ and agents $l \in \mathcal{A}_{c,\gamma}^{[i]}$ for which the safety respectively velocity constraints are not satisfied. Furthermore, $\varepsilon_{d,l,(k+j|k)}^{[i]} \geq 0$ and $\varepsilon_{v,(k+j|k)}^{[i]} \geq 0$ reflect the amount of constraint violation, i.e.,

$$\varepsilon_{d,l,(k+j|k)}^{[i]} = d_{safe,l,(k+j|k)}^{[i]} - \tilde{d}_{l,(k+j|k)}^{[i]}, \quad \varepsilon_{v,(k+j|k)}^{[i]} = \tilde{v}_{(k+j|k)}^{[i]} - \overline{v}_{(k+j|k)}^{[i]}. \quad (9.24)$$

This randomization approach is summarized in Algorithm 1.

Algorithm 1 Randomization using soft constraints.

1: $\tilde{u}_{(\cdot|k)}^{[i]*} \leftarrow u_{(\cdot|k)}^{[i]*}$, $\tilde{c}_{min}^{[i]} \leftarrow \text{COST}(x_0^{[i]}, u_{(\cdot|k)}^{[i]*})$ ▷ Initial solution
2: **for** $n := 1$ to $N_{samples}$ **do**
3:     $\tilde{u}_{(\cdot|k)}^{[i]} \leftarrow \mathcal{N}(u_{(\cdot|k)}^{[i]*}, U^{[i]*} - u_{(\cdot|k)}^{[i]*}u_{(\cdot|k)}^{[i]*,T})$ ▷ Draw sample from distribution
4:     $\tilde{c}^{[i]} \leftarrow \text{COST}(x_0^{[i]}, \tilde{u}_{(\cdot|k)}^{[i]})$ ▷ Sample cost
5:     **if** $\tilde{c}^{[i]} < \tilde{c}_{min}^{[i]}$ **then**
6:       $\tilde{c}_{min}^{[i]} \leftarrow \tilde{c}^{[i]}$, $\tilde{u}_{(\cdot|k)}^{[i]*} \leftarrow \tilde{u}_{(\cdot|k)}^{[i]}$ ▷ Take sample if cost is lower
7:     **end if**
8: **end for** ▷ Optimal solution is $\tilde{u}_{(\cdot|k)}^{[i]*}$
9:
10: **function** COST($x_0^{[i]}, \tilde{u}_{(\cdot|k)}^{[i]}$)
11:     $\tilde{c}^{[i]} \leftarrow \inf$ ▷ Initialize cost
12:     **if** hard constraints satisfied **then**
13:       $\tilde{c}^{[i]} \leftarrow J^{[i]}(x_0^{[i]}, \tilde{u}_{(\cdot|k)}^{[i]}) + J_{viol}^{[i]}(x_0^{[i]}, \tilde{u}_{(\cdot|k)}^{[i]})$ ▷ Total cost w/ violation cost
14:     **end if**
15:     **return** $\tilde{c}^{[i]}$
16: **end function**

9.5 Feasibility and Optimality

Having shown how to formulate and solve the distributed OCP, we finally have to discuss local and global feasibility as well as optimality of our approach. In the remainder of this section, we will therefore conduct a brief analysis on these topics.

Proposition 9.5.1 (Local Feasibility) *For every agent $i \in \mathcal{A}$, a local solution to an accomplishable conflict resolution problem can be found if the prediction horizon is of sufficient length.*

Proof (Sketch) A feasible local solution is obtained if SDP relaxation solves the original problem or randomization (adopting soft constraints) determines a feasible solution. To ensure that an a priori accomplishable scenario does not become infeasible, the prediction horizon has to be of sufficient length to allow every agent to act sufficiently early. Therefore, we ensure that the prediction horizon at least covers the coordinate set $\mathcal{S}_{cb}^{[i]}$ when we approach the brake safe distance.

If the conflict resolution problem is accomplishable, optimization will be able to determine a feasible solution as it is aware of the entire conflict region before entering and while passing it.

Proposition 9.5.2 (Global Feasibility) *Given that agent priorities are time-invariant, then an accomplishable conflict resolution problem is globally feasible if there exists a feasible local solution for every agent $i \in \mathcal{A}$.*

Proof (*Sketch*) In the distributed OCP with time-invariant priorities, the agent with the highest priority is in favor of solving its local OCP without any safety constraints. As such, its input and state trajectories will converge for the given scenario. Then, the agent having the next lower priority will adapt to the previous agent by imposing safety constraints on the first agent. When the first agent's solution has converged, the second agent's solution will converge as well. We can continue this consideration for the following agents with the next lower priorities. As such, the distributed OCPs will converge sequentially such that the scenario is finally globally feasible.

Proposition 9.5.3 (Optimality) *The prioritized distributed OCP always solves the centralized OCP in a suboptimal way, except in the trivial case when none of the safety constraints become active. The prioritized distributed OCP is solved globally optimal if a minimizer can be found for every local subproblem which fulfills the rank-1 condition* (9.20).

Proof (*Sketch*) In the centralized problem, safety constraints are imposed pairwise on potentially conflicting agents which might result in a mutual reaction of both agents to avoid a collision. In the prioritized distributed OCP, safety constraints are only imposed on the agent with lower priority which implies that only that agent is able to react on the high-priority agent. Consequently, suboptimal solutions are obtained unless we have the trivial case that none of the safety constraints become active.

For the prioritized distributed OCP, the global optimum is obtained if a global minimizer can be found for every local subproblem. A global minimizer of a local subproblem satisfies the rank-1 condition for the Schur complement matrix $\mathcal{S}^{[i]}$. If this condition is not fulfilled for at least one agent, the prioritized distributed OCP is solved in a suboptimal way.

In the context of optimality, the reader should be aware that we aim at solving a non-convex optimization problem which is indeed a challenging task. Therefore, solving the optimization problem in a suboptimal but computationally tractable way is a common approach to obtain a solution of such kind of optimization problems.

9.6 Simulation Study

In this section, the proposed distributed optimization approach is evaluated in a comprehensive simulation study focusing on an urban intersection scenario. In the following, we will outline the scenario setup first before finally discussing simulation results.

9.6.1 Intersection Scenario

For evaluation purposes, an urban four-way intersection scenario including four agents, as illustrated in Fig. 9.2, is investigated. Particularly, agent 2 (blue) and agent 3 (red) are driving straight in opposite directions, while agent 1 (green) is turning right and agent 4 (cyan) is turning left. As such, agents 1, 2, and 4 are in conflict as they share the same target lane. Moreover, agent 4 crosses the straight path of agent 3 while turning left.

For the simulation study, every agent is parametrized equally; i.e., every agent has the same length $L^{[i]} = 4.87$ m, width $W^{[i]} = 1.85$ m as well as the same dynamic time constant $T_{a_x}^{[i]} = 0.3$ s which reflects the agent's ability to accelerate respectively to decelerate. In our scenario, three agents have the same target lane; i.e., after passing the collision point they are all driving in the same lane. Obviously, in case that these agents have different desired speeds after the collision point, we require additional constraints that prevent the agents from colliding in the same lane. Although such

Fig. 9.2 Intersection scenario: two agents going straight, one agent turning left, one agent turning right

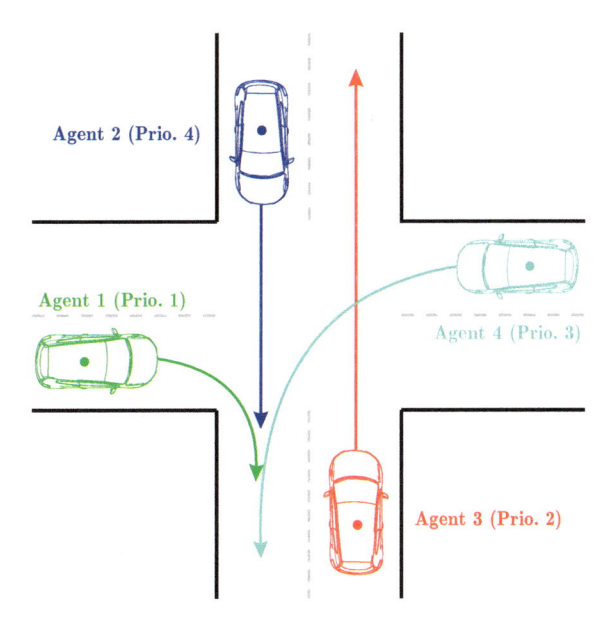

Table 9.1 Scenario setup: initial conditions

	Init. position $s_0^{[i]}$ (m)	Init. velocity $v_0^{[i]}$ (m/s)	Priority (−)
Agent 1 (green)	−50.5	13.9	1
Agent 2 (blue)	−102.8	13.9	4
Agent 3 (red)	−70.4	13.9	2
Agent 4 (cyan)	−78.5	13.9	3

constraints can easily be integrated in our approach, we omit these constraints as such a scenario is not in the scope of this contribution. Moreover, we assume that another control system like adaptive cruise control might take over control to avoid rear-end collisions and keep a safe distance in the same lane.

The required safety distances between the agents have been determined based on the vehicle geometry and the relative orientation of the agents in the scenario; i.e., for turning vehicles, we need a larger safety margin than for straight crossing vehicles. The local MPC controllers have a sample time of $T_s = 0.25$ s and a horizon length of $H_u = H_p = 20$, thus exploiting a preview time of 5 s. Controller parameterization is equal for all agents using the following weighting coefficients: $Q^{[i]} = 1$, $R^{[i]} = 2$, $S^{[i]} = 2$, $\rho_v = 10^3$, $\rho_d = 10^6$. Furthermore, the minimum longitudinal acceleration and maximum longitudinal acceleration have been chosen as $\underline{u}^{[i]} = -9\,\text{m/s}^2$ and $\overline{u}^{[i]} = 5\,\text{m/s}^2$. As previously outlined, we impose a lower velocity bound of zero to avoid predictions that indicate that an agent is driving backward, while the upper bound is set to 110% of the desired speed, i.e., $\overline{v}^{[i]}_{(k+j|k)} = 1.1 \cdot v_{ref,(k+j|k)}$ for $j \in \{1, \ldots, H_p\}$. On a straight, the desired speed is equal to the speed limit, while in a curve the desired speed is set to the minimum of the speed limit and a curve speed that does not cause absolute lateral accelerations larger than $3\,\text{m/s}^2$. For solving the local optimization problem, we apply SDPA [30] as SDP solver, apply a threshold of $\sigma_{2,thld} = 10^{-4}$ for the second largest singular value, and draw 100 samples for randomization. The simulation platform is an Intel i5 2.7 GHz laptop with MATLAB/Simulink R2015b. By adopting the time to react criteria for prioritization, see Sect. 9.4.3, the initial conditions of the simulation scenario can be summarized as illustrated in Table 9.1. Recalling the definition in Sect. 9.2.1, the origin of the agent's position coordinate frame refers to the first collision point with another agent.

9.6.2 Discussion of Results

The results of the intersection scenario are illustrated in Fig. 9.3. The first row shows (from left to right) a basic scenario overview, the second largest singular value of the Schur complement matrix for every agent as well as the solver exitflags of the local optimization problems. Particularly, for an exitflag of 0, SDR has solved the original non-convex problem; for an exitflag of 1, randomization has been necessary

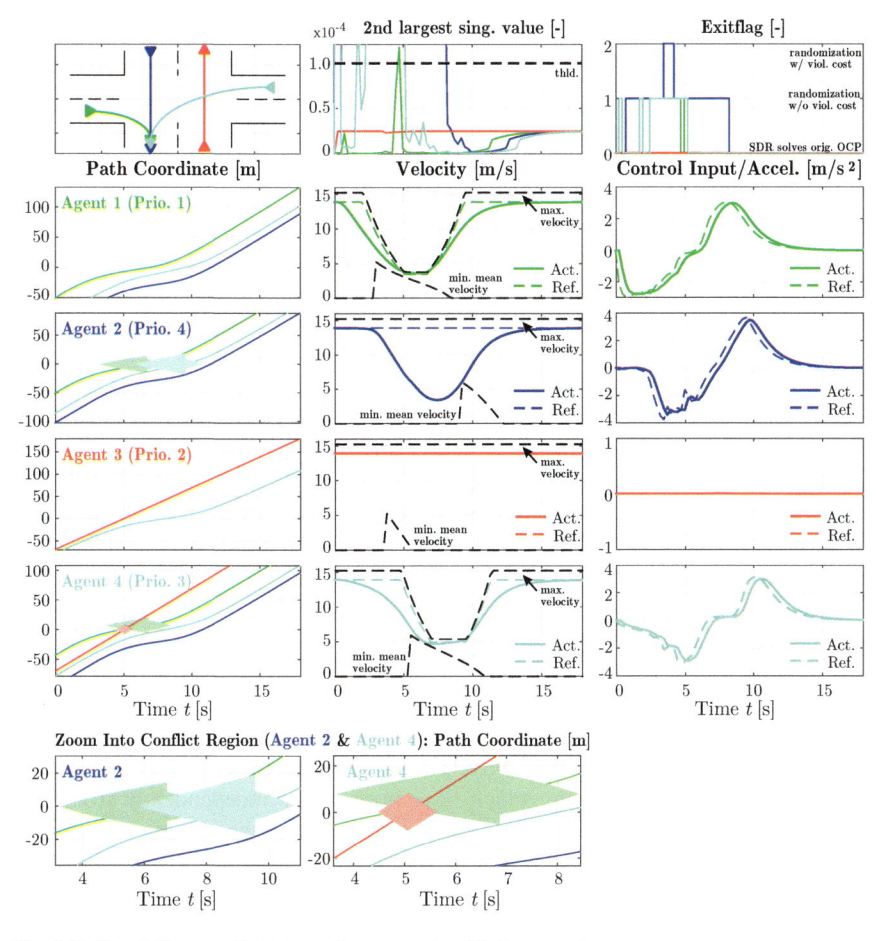

9 Automation of Road Intersections Using Distributed Model Predictive Control

Fig. 9.3 Simulation results: intersection scenario with four agents

without any violation cost; and for an exitflag 2, randomization has been applied with a violation cost larger than zero. In the next four rows, we illustrate the agents' path coordinate, velocity, and acceleration (from left to right). In the first column, we show the path coordinate trajectory of agent i along with the trajectory of every other agent l that might potentially be in conflict with agent i in the reference frame of agent i. Again, a path coordinate of zero refers to the first collision point in the reference frame of agent i. In case that a safety constraint is required, a colored polygon around the trajectory of agent l illustrates the area that must not be entered by the trajectory of agent i in order to satisfy safety constraints. In the second column, we outline the actual velocity (colored solid line) along with its reference (colored dashed line) originating from a high-level planning algorithm as well as the maximum and minimum mean velocity (black dashed lines). In the rightmost column, we have plotted the reference acceleration, i.e., the control input, resulting from the local

MPC controllers (dashed line) along with the actual longitudinal acceleration (solid line). As demanded accelerations never get close to their bounds, these are left out for reasons of clarity. In the last row, we provide a closer insight into the conflict region for agents 2 and 4 to better assess constraint satisfaction.

Analyzing the results in Fig. 9.3, we can recognize that agent 1, having the highest priority, passes the intersection without taking care of any other agent. As such, agent 1 solely follows its prescribed velocity reference trajectory without violating any bounds and turns right when entering the intersection. Agent 3, having the next lower priority, is not in conflict with agent 1 and is thus able to pass the intersection without any safety constraints almost simultaneously with agent 1. When agent 4 is turning left, it gives right-of-way to agents 1 and 3 by passing the corresponding collision point after these agents with sufficient safety distance. Finally, agent 2 passes its collision point after agents 1 and 4 maintaining its prescribed safety distance. To prove constraint satisfaction for agent 2 and agent 4, the reader is referred to the bottom row in Fig. 9.3 which depicts a closer insight into the corresponding conflict region. The attentive reader might have recognized that the size of the safety interval around the trajectory of the conflicting agent decreases after that agent has passed the joint collision point. According to Sect. 9.3, we would like to recall that we reduce the required safety distance after one of the conflicting agents has passed the respective collision point. Besides satisfying the safety constraints, it is obvious that both agents try to remain as close as possible to the other agents when passing the collision point to avoid slowing down. This behavior corresponds to the objective to minimize the difference between the desired and actual velocity. Finally, it can be proven that safety constraints are not only satisfied in discrete-time but also in continuous-time. Apparently, the safety polygons which have been drawn independently from the sampling time are never intersected by the trajectories of the conflicting agents.

Finally, we would like to focus on the ability of our approach to solve the distributed OCP and on its computational complexity. Referring to the second largest singular value of the Schur complement matrix respectively the exitflag of the solver, we can recognize for agent 3 that SDR solves the original problem during the entire maneuver as there are no non-convex constraints. For agent 1, not having any non-convex constraints either, there is just one time step in which randomization is applied. During that time step, the second largest singular value slightly exceeds the user-defined threshold which is rather a numerical effect. For agent 4, randomization without utilizing any violation cost is frequently required during the first 5.5s of the maneuver when safety constraints are imposed on the agent. Eventually considering agent 2, having the lowest priority, it can be recognized that randomization is applied for the first 8s with occasionally utilizing the violation cost to obtain a feasible solution. Nevertheless, utilizing the violation cost over the prediction horizon does not lead to any noticeable constraint violations. Herewith, we can conclude that our approach is able to find a solution for every local OCP such that we finally converge to a distributed solution that satisfies all our control objectives.

The previous observations that have been made in terms of solver flags are also reflected by the required mean and maximum computation time as summarized in Table 9.2. To obtain a better insight, we broke down these times into the mean and

Table 9.2 Mean and max. computation time for every agent

		SDR (ms)	Randomization (ms)	Total (ms)
Agent 1 (green)	Mean	45.1	0.2	45.5
	Max.	63.5	8.6	63.8
Agent 2 (blue)	Mean	78.0	4.0	82.1
	Max.	146.1	9.4	155.0
Agent 3 (red)	Mean	45.0	0.1	45.7
	Max.	52.3	0.2	63.2
Agent 4 (cyan)	Mean	63.5	1.7	65.4
	Max.	117.5	9.3	126.3

maximum time for solving the SDR and conducting randomization. The reader has to be aware that we have determined these execution times separately for SDR, randomization, and the total computation time such that they do not necessarily sum up. Finally, with a sample time of 250 ms and a maximum total execution time of 155.0 ms we can conclude that we are always able to solve the distributed control problem during a sampling step which is a very satisfying result given that we have to solve a non-convex problem. Nevertheless, further improvement of computation time seems to be required for experimental evaluation purposes. The attentive reader might eventually question why the sampling time cannot be reduced any further when there is still a gap of approximately 100ms between the maximum computation time and the sampling time. A reduction of the sampling time would then either require to enlarge the prediction horizon or to increase the minimum mean velocity to maintain the required preview for collision avoidance. In the former case, computation time would rise due to an increased dimensionality of the optimization problem such that a solution could no longer be obtained in real time. The latter case might result in an infeasible scenario when the minimum mean velocity constraint can no longer be satisfied.

9.7 Conclusions and Future Work

Intersection automation leveraging V2X communication offers significant potential to improve traffic efficiency and throughput. However, the complexity that arises with such kind of scenarios is quite challenging as we require a resilient, robust, and scalable control scheme which has to deal with non-convex optimization problems and which allows for computing control actions efficiently. To accommodate these challenges, we have introduced a distributed MPC approach which relies on V2V communication for exchanging information with other agents. The proposed method allows multiple agents to enter the intersection simultaneously instead of blocking it exclusively. Moreover, the distributed optimal control problem is fully parallelized without requiring any nested iterations. This parallelized approach is inevitable for an implementation in an actual vehicle as nested iterations might require additional inter-vehicle communication effort for a solving a single distributed optimization problem at time k. In a simulation study, we have investigated an urban intersection scenario with four agents. The results show that our distributed control concept is capable of finding a distributed solution of the OCP that meets all our objectives. Furthermore, computation times indicate that our approach shows promising potential to be implemented on a rapid control prototyping hardware and as such in an actual vehicle.

Regarding future work, we will analyze how computation times can further be reduced. Furthermore, it will be investigated how constraint priorities can be negotiated among the agents either in a distributed hierarchical framework or as a part of the distributed OCP itself. Moreover, including more accurate vehicle models, eventually also lateral vehicle motion, will be considered as part of our future research work. Eventually, we are envisioning to evaluate the proposed approach in an experimental setup. In that context, additional challenges that arise, e.g., with uncertainties in the communication channel or in vehicle localization have to be addressed as well.

Acknowledgements This contribution extends the previous publication [20] which is under copyright of the International Federation of Automatic Control (IFAC), 2017. Parts of this contribution (modeling, description, and implementation of the control and optimization problem) have been reused with the permission of IFAC which is acknowledged with high appreciation. This chapter, though, provides a more comprehensive overview of the applied algorithms and an extensive analysis of simulation results in a different intersection scenario.

References

1. Chen, L., Englund, C.: Cooperative intersection management: a survey. IEEE Trans. Intell. Transp. Syst. **17**(2), 570–586 (2016)
2. Wymeersch, H., de Campos, G.R., Falcone, P., Svensson, L., Ström, E.G.: Challenges for cooperative its: improving road safety through the integration of wireless communications, control, and positioning. In: International Conference on Computing, Networking and Communications, pp. 573–578 (2015)
3. Kamal, M.A.S., i. Imura, J., Hayakawa, T., Ohata, A., Aihara, K.: A vehicle-intersection coordination scheme for smooth flows of traffic without using traffic lights. IEEE Trans. Intell. Transp. Syst. **16**(3), 1136–1147 (2015)
4. Murgovski, N., de Campos, G.R., Sjberg, J.: Convex modeling of conflict resolution at traffic intersections. In: IEEE Conference on Decision and Control, pp. 4708–4713 (2015)
5. Quinlan, M., Au, T.C., Zhu, J., Stiurca, N., Stone, P.: Bringing simulation to life: a mixed reality autonomous intersection. In: Proceedings of IROS 2010-IEEE/RSJ International Conference on Intelligent Robots and Systems (IROS 2010) (2010)
6. de Campos, G.R., Falcone, P., Wymeersch, H., Hult, R., Sjoberg, J.: Cooperative receding horizon conflict resolution at traffic intersections. In: IEEE Conference on Decision and Control, pp. 2932–2937 (2014)
7. Hafner, M.R., Cunningham, D., Caminiti, L., Vecchio, D.D.: Cooperative collision avoidance at intersections: algorithms and experiments. IEEE Trans. Intell. Transp. Syst. **14**(3), 1162–1175 (2013)

8. Qian, X., Gregoire, J., de La Fortelle, A., Moutarde, F.: Decentralized model predictive control for smooth coordination of automated vehicles at intersection. In: European Control Conference, pp. 3452–3458 (2015)
9. Hult, R., Campos, G.R., Falcone, P., Wymeersch, H.: An approximate solution to the optimal coordination problem for autonomous vehicles at intersections. In: IEEE American Control Conference, pp. 763–768 (2015)
10. Kim, K.D., Kumar, P.R.: An mpc-based approach to provable system-wide safety and liveness of autonomous ground traffic. IEEE Trans. Autom. Control **59**(12), 3341–3356 (2014)
11. Gregoire, J., Bonnabel, S., de La Fortelle, A.: Optimal cooperative motion planning for vehicles at intersections. In: Accurate Positioning and Mapping for Intelligent Vehicles, IEEE Intelligent Vehicles Symposium (2012)
12. Hafner, M.R., Vecchio, D.D.: Computational tools for the safety control of a class of piecewise continuous systems with imperfect information on a partial order. SIAM J. Control Optim. **49**(6), 2463–2493 (2011)
13. Dresner, K., Stone, P.: A multiagent approach to autonomous intersection management. J. Artif. Intell. Res. **31**(1), 591–656 (2008)
14. Kowshik, H., Caveney, D., Kumar, P.R.: Provable systemwide safety in intelligent intersections. IEEE Trans. Veh. Technol. **60**(3), 804–818 (2011)
15. Ahn, H., Colombo, A., Vecchio, D.D.: Supervisory control for intersection collision avoidance in the presence of uncontrolled vehicles. In: 2014 American Control Conference, pp. 867–873 (2014)
16. Bruni, L., Colombo, A., Vecchio, D.D.: Robust multi-agent collision avoidance through scheduling. In: IEEE Conference on Decision and Control, pp. 3944–3950 (2013)
17. Colombo, A., Vecchio, D.D.: Least restrictive supervisors for intersection collision avoidance: a scheduling approach. IEEE Transa. Autom. Control **60**(6), 1515–1527 (2015)
18. Kim, K.D.: Collision free autonomous ground traffic: a model predictive control approach. In: ACM/IEEE International Conference on Cyber-Physical Systems (ICCPS), pp. 51–60 (2013)
19. Medina, A.I.M., Van De Wouw, N., Nijmeijer, H.: Automation of a T-intersection using virtual platoons of cooperative autonomous vehicles. In: IEEE International Conference on Intelligent Transportation Systems, pp. 1696–1701 (2015)
20. Katriniok, A., Kleibaum, P., Joševski, M.: Distributed model predictive control for intersection automation using a parallelized optimization approach. In: IFAC World Congress (2017)
21. Frazzoli, E., Mao, Z.H., Oh, J.H., Feron, E.: Resolution of conflicts involving many aircraft via semidefinite programming. J. Guidance Control Dyn. **24**(1), 79–86 (2001)
22. Johansson, B., Keviczky, T., Johansson, M., Johansson, K.H.: Subgradient methods and consensus algorithms for solving convex optimization problems. In: IEEE Conference on Decision and Control, pp. 4185–4190 (2008)
23. Nedic, A., Ozdaglar, A.: Distributed subgradient methods for multi-agent optimization. IEEE Trans. Autom. Control **54**(1), 48–61 (2009)
24. Margellos, K., Falsone, A., Garatti, S., Prandini, M.: Proximal minimization based distributed convex optimization. In: IEEE American Control Conference, pp. 2466–2471 (2016)
25. Wei, E., Ozdaglar, A.: On the o(1 = k) convergence of asynchronous distributed alternating direction method of multipliers. In: IEEE Global Conference on Signal and Information Processing, pp. 551–554 (2013)
26. de Campos, R., Falcone, P., Sjöberg, J.: Traffic safety at intersections: a priority based approach for cooperative collision avoidance. In: International Symposium on Future Active Safety Technology Towards Zero Traffic Accidents, pp. 9–15 (2015)
27. Boyd, S., Vandenberghe, L.: Semidefinite Programming Relaxations of Non-Convex Problems in Control and Combinatorial Optimization, pp. 279–287. Springer, Boston (1997)
28. Cheng, Y., Haghighat, S., Cairano, S.D.: Robust dual control MPC with application to soft-landing control. In: IEEE American Control Conference, pp. 3862–3867 (2015)
29. d'Aspremont, A., Boyd, S.: Relaxations and Randomized Methods for Nonconvex QCQPs. Stanford University, Stanford (2003)
30. Fujisawa, K., Nakata, K., Yamashita, M., Fukuda, M.: SDPA project: solving large-scale semidefinite programs. J. Oper. Res. Soc. Jap. **50**(4), 278–298 (2007)

Chapter 10
MPDM: Multi-policy Decision-Making from Autonomous Driving to Social Robot Navigation

Alex G. Cunningham, Enric Galceran, Dhanvin Mehta, Gonzalo Ferrer, Ryan M. Eustice and Edwin Olson

Abstract This chapter presents multi-policy decision-making (MPDM): a novel approach to navigating in dynamic multi-agent environments. Rather than planning the trajectory of the robot explicitly, the planning process selects one of a set of closed-loop behaviors whose utility can be predicted through forward simulation that captures the complex interactions between the actions of these agents. These polices capture different high-level behavior and intentions, such as driving along a lane, turning at an intersection, or following pedestrians. We present two different scenarios where MPDM has been applied successfully: an autonomous driving environment models vehicle behavior for both our vehicle and nearby vehicles and a social environment, where multiple agents or pedestrians configure a dynamic environment for autonomous robot navigation. We present extensive validation for MPDM on both scenarios, using simulated and real-world experiments.

Alex G. Cunningham and Enric Galceran have contributed equally to this work.

A. G. Cunningham
Toyota Research Institute, 2311 Green Rd, Ann Arbor, MI 48105, USA
e-mail: alex.cunningham@tri.global

E. Galceran
Autonomous Systems Lab, Institute of Robotics and Intelligent Systems, ETH Zurich,
Leonhardstrasse 21, 8092 Zurich, Switzerland
e-mail: enricg@ethz.ch

D. Mehta (✉) · G. Ferrer · E. Olson
Department of Computer Science and Engineering, University of Michigan,
2260 Hayward St, Ann Arbor, MI 48109, USA
e-mail: dhanvinm@umich.edu

G. Ferrer
e-mail: gferrerm@umich.edu

E. Olson
e-mail: ebolson@umich.edu

R. M. Eustice
Department of Naval Architecture and Marine Engineering, University of Michigan,
2600 Draper Dr, Ann Arbor, MI 48109, USA
e-mail: eustice@umich.edu

© Springer International Publishing AG, part of Springer Nature 2019
H. Waschl et al. (eds.), *Control Strategies for Advanced Driver Assistance Systems and Autonomous Driving Functions*, Lecture Notes in Control and Information Sciences 476, https://doi.org/10.1007/978-3-319-91569-2_10

10.1 Introduction

Decision-making in dynamic multi-agents environments is challenging due to the uncertainty associated with estimating and predicting future scenarios arising from the complex and tightly coupled interactions between agents. Sensor noise, action execution uncertainty, tracking data association errors, etc., make this problem harder.

The robot's plan must consider the uncertainty on the continuous state of nearby agents and, especially, over their potential discrete intentions, such as turning at an intersection or changing lanes (Fig. 10.1). Our goal is to correctly handle uncertainty and prediction, while calculating a solution within a time budget for online execution, as a direct application for ADF. The robustness on prediction though MPDM is also useful to facilitate the driving experience and its potential implementation in ADAS.

This chapter describes a planning framework called multi-policy decision-making (MPDM), which poses planning as a discrete-valued decision problem over a library of hand-engineered closed-loop behavioral policies. We treat the underlying behavioral policies, roughly equivalent to control laws as black boxes whose outputs can be predicted using a forward simulation.

For any particular situation, our approach chooses the best policy online by simulating the likely future outcomes of each policy over a time horizon. The robot does not compute a nominal trajectory; it simply "elects" a particular closed-loop behavior until the planning cycle runs again. Dynamically switching between candidate policies allows the robot to adapt to different situations that are likely to arise.

MPDM can be applied to a variety of domains, ranging from autonomous driving to mobile indoor robots. It is a powerful framework in that it allows relatively complex behaviors to be derived from a relatively small set of underlying policies.

This chapter is structured as follows: in the following Sect. 10.3, we will present the formulation required to obtain the MPDM. The next two sections describe the authors application of MPDM to a number of real-world systems, including both how the basic formulation of MPDM was adapted from one setting to another and how the underlying policies were developed. In particular, an autonomous driving

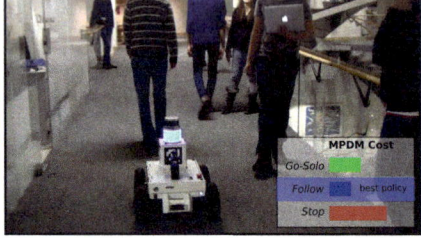

Fig. 10.1 *Left* Our multi-policy approach factors the actions of the egovehicle and traffic vehicles into a set of policies that capture common behaviors like lane following, lane changing, or turning, from [1] © 2017 Springer, reproduced with permission. *Right* Multi-policy applied to a different scenario, a social environment, from [2] © 2016 IEEE, reproduced with permission

scenario (Sect. 10.4) based on our previous publications, see [1, 3, 4] for more details. A second scenario, a social environment (Sect. 10.5) borrows from our IROS publication [2].

10.2 Related Work

10.2.1 Related Work on Behavioral Prediction

Despite the probabilistic nature of the anticipation problem, several methods in the literature assume no uncertainty on the future states of other participants [5–7]. Such an approach could be justified in a scenario where vehicles broadcast their intentions over some communications channel, but it is an unrealistic assumption otherwise.

Some approaches assume a dynamic model of the obstacle and propagate its state using standard filtering techniques such as the extended Kalman filter [8, 9]. Despite providing rigorous probabilistic estimates over an obstacle's future states, these methods often perform poorly when dealing with nonlinearities in the assumed dynamics model and the multimodalities induced by discrete decisions (e.g., continuing straight, merging, or passing). Some researchers have explored using Gaussian mixture models (GMM) to account for nonlinearities and multiple discrete decisions [10, 11]; however, these approaches do not consider the history of previous states of the target object, assigning an equal likelihood to each discrete hypothesis and leading to a conservative estimate.

Dynamic Bayesian networks have been also utilized for behavioral anticipation [12]. In [13] is proposed a hierarchical dynamic Bayesian network where some of the models on the network are learned from observations using an EM approach.

A common anticipation strategy in autonomous driving used by, for example, [14, 15], or [16], consists of computing the possible goals of a target vehicle by planning from its standpoint, accounting for its current state. This strategy is similar to our factorization of potential driving behavior into a set of policies, but lacks closed-loop simulation of vehicle interactions.

Gaussian process (GP) regression has been utilized to learn typical motion patterns for classification and prediction of agent trajectories [17–19], particularly in autonomous driving [20–22]. In more recent work, [23] use inverse reinforcement learning to learn driving styles from trajectory demonstrations in terms of engineered features. They then use trajectory optimization to generate trajectories for their autonomous vehicle that resembles the learned driving styles. Nonetheless, these methods require the collection of training data to reflect the many possible motion patterns the system may encounter, which can be time-consuming. For instance, a lane change motion pattern learned in urban roads will not be representative of the same maneuver performed at higher speeds on the highway. In this paper, we focus instead on hand-engineered policies.

10.2.2 Related Work on Decision-Making

Early instances of decision-making systems for autonomous vehicles capable of handling urban traffic situations stem from the 2007 DARPA Urban Challenge [24]. In that event, participants tackled decision-making using a variety of solutions ranging from FSM [25] and decision trees [26] to several heuristics [27]. However, these approaches were tailored for specific and simplified situations and were, even according to their authors, "not robust to a varied world" [27].

More recent approaches have addressed the decision-making problem for autonomous driving through the lens of trajectory optimization [16, 28–30].

However, these methods do not model the closed-loop interactions between vehicles, failing to reason about their potential outcomes.

Partially observable Markov decision processes (POMDP) offer a theoretically grounded framework to incorporate these interactions in the planning process; however, solvers [31–33] often have difficulty scaling computationally to real-world scenarios. The POMDP model provides a mathematically rigorous formalization of the decision-making problem in dynamic, uncertain scenarios such as autonomous driving. Unfortunately, finding an optimal solution to most POMDP is intractable [34, 35]. A variety of general POMDP solvers exist in the literature that seek to approximate the solution [31–33, 36]. Although these methods typically require computation times on the order of several hours for problems with even small state, observation, and action spaces compared to real-world scenarios [37], there has been some recent progress that exploits GPU parallelization [38].

However, some researchers have proposed approximate solutions to the POMDP formulation to tackle decision-making in autonomous driving scenarios. Reference [39] proposed a point-based MDP for single-lane driving and merging, and [40] applied a POMDP formulation to handle highway lane changes. An MDP formulation was employed by [41] for highway driving; similarly to our *policies*, they utilize *behaviors* that react to other objects. The POMDP approach of [42] considers partial observability of road users' intentions, while [43] solve a POMDP in continuous state space reasoning about potentially hidden objects and observation uncertainty, considering the interactions of road users.

The idea of assuming finite sets of policies to speed up planning has appeared previously [41, 43–46]. Similarly, we propose to exploit domain knowledge from autonomous driving to design a set of policies that are readily available at planning time.

10.2.3 Related Work on Social Navigation

In a simulated environment, van den Berg et al. [47] proposed a multi-agent navigation technique using *velocity obstacles* that guarantee a collision-free solution assuming a fully observable world. From the computer graphics community, Guy et al. [48] extended this work using *finite-time velocity obstacles* to provide a locally collision-free solution that was less conservative as compared to [47]. However, the main drawback of these methods is that they are sensitive to imperfect state estimates and makes strong assumptions that may not hold in the real world.

Several approaches attempt to navigate in social environments by traversing a potential field (PF) [49] generated by a set of pedestrians [50–52]. Huang et al. [53] used visual information to build a PF to navigate. In the field of neuroscience, Helbing and Molnár [54] proposed the social force model, a kind of PF approach that describes the interactions between pedestrians in motion.

Unfortunately, PF approaches have some limitations, such as local minima or oscillation under certain configurations [55]. These limitations can be overcome to a certain degree by using a global information plan to avoid local minima [56]. We use this same idea in our method by assuming that a global planner provides reachable goals, i.e., there is a straight line connection to those positions ensuring feasibility in the absence of other agents.

Inverse reinforcement learning-based approaches [57–60] can provide good solutions by predicting social environments and planning through them. However, their effectiveness is limited by the training scenarios considered which might not be a representative set of the diverse situations that may arise in the real world.

An alternative approach looks for a pedestrian leader to follow, thus delegating the responsibility of finding a path to the leader, such as the works of [61–63]. In this work, *Follow* becomes one of the policies that the robot can choose to execute as an alternate policy to navigating.

Some approaches [64–67] plan over the predicted trajectories of other agents. However, predicting the behavior of pedestrians is challenging and the underlying planner must be robust to prediction errors.

POMDPs provide a principled approach to deal with uncertainty, but they quickly become intractable. Foka et al. [68] used POMDPs for robot navigation in museums. Cunningham et al. [3] show that, by introducing a number of approximations (in particular, constraining the policy to be one of a finite set of known policies), that the POMDP can be solved using MPDM. In their original paper, they use a small set of lane-changing policies; in this work, we explore an indoor setting in which the number and complexity of candidate policies are much higher [2].

10.3 Problem Formulation

Partially observable Markov decision processes (POMDP) provides a mathematically rigorous formalization of the decision-making problem in dynamic, uncertain scenarios. We initially formulate this problem as a full POMDP which we then approximate by exploiting domain knowledge to reformulate the problem as a discrete decision over a small set of high-level policies for the robot.

Let V denote the set of agents near the robot including the robot. At time t, an agent $v \in V$ can take an action $a_t^v \in A^v$ to transition from state $x_t^v \in X^v$ to x_{t+1}^v. As a notational convenience, let $x_t \in X$ include all state variables x_t^v for all agents at time t, and similarly let $a_t \in A$ be the actions of all agents.

We model the agent dynamics with a conditional probability function capturing the dependence of the dynamics on the states and actions of all the agents in the neighborhood.

$$T(x_t^v, a_t, x_{t+1}) = p(x_{t+1}^v | x_t, a_t). \tag{10.1}$$

Similarly, we model observation uncertainty as

$$Z(x_t, z_t^v) = p(z_t^v | x_t), \tag{10.2}$$

where $z_t^v \in Z^v$ is the observation made by agent v at time t, and $z_t \in Z$ is the vector of all sensor observations made by all agents. These observations are provided by the perception module to the robot (see Fig. 10.2). For the rest of the agents considered during planning, we transform the observations into each agent's coordinate frame, considering the robot's state as an observation.

The robot's goal is to find an optimal policy π^* that maximizes the expected sum of rewards over a given decision horizon H, where a policy is a mapping $\pi : X \times Z^v \to A^v$ that yields an action from the current MAP estimate of the state and an observation:

$$\pi^* = \arg\max_\pi E\left[\sum_{t=t_0}^{H} R(x_t, \pi(x_t, z_t^v))\right], \tag{10.3}$$

where $R(\cdot)$ is a real-valued reward function $R : X \times A \to \mathbb{R}$. The evolution of $p(x_t)$ over time is governed by

$$p(x_{t+1}) = \iiint_{XZA} p(x_{t+1}|x_t, a_t) p(a_t|x_t, z_t) p(z_t|x_t) p(x_t) \, da_t \, dz_t \, dx_t. \tag{10.4}$$

However, modeled agents can still react to nearby agents via z_t^v. Thus, the joint density for a single agent v can be written as

$$p^v(x_t^v, x_{t+1}^v, z_t^v, a_t^v) = p(x_{t+1}^v | x_t^v, a_t^v) p(a_t^v | x_t^v, z_t^v) p(z_t^v | x_t^v) p(x_t^v), \tag{10.5}$$

and assuming independent agent actions leads to

$$p(x_{t+1}) = \prod_{v \in V} \iiint_{X^v Z^v A^v} p^v(x_t^v, x_{t+1}^v, z_t^v, a_t^v) \, da_t^v \, dz_t^v \, dx_t^v. \tag{10.6}$$

Despite the independence assumption, marginalizing over the large state, observation, and action spaces in Eq. 10.6 is still too expensive. A possible approximation to speed up the process, commonly used by general POMDP solvers [31, 36], is to solve Eq. 10.3 by drawing samples from $p(x_t)$. However, sampling over the full probability space with random walks yields a large number of low probability samples, such as those with agents not abiding by traffic rules. Our proposed approach samples more strategically from high likelihood scenarios to ensure computational tractability.

10.3.1 The Multi-Policy Approximation

The key idea we leverage is that, rather than plan nominal trajectories, we can think of behavior as emerging from choosing closed-loop policies. For instance, in indoor social environments, the robot can plan in terms of following or stopping. In the vast majority of traffic situations, traffic participants behave in a regular, predictable manner, following traffic rules. In general, these policies are hand-engineered, and thoroughly evaluated, discarding them when they result in infeasible trajectories. Typical behaviors that conform to these rules can greatly limit the action space to be considered and provide a natural way to capture closed-loop interactions. Thus, we can structure the decision process to reason over a limited space of closed-loop policies for both the robot and other agents.

Closed-loop policies[1] allow approximation of agent dynamics including their interactions and observation models from Sect. 10.3 through deterministic, coupled forward simulation of multiple agents with their assigned policies. Therefore, we can evaluate the consequences of our decisions over available policies (for both the robot and other agents), without needing to evaluate for every control input of every agent.

This assumption does not preclude our system from handling situations where reaction time is key, as we engineer all policies to produce robot behavior that seeks safety at all times.

More formally, let Π be a discrete set of policies π_i, where each policy is a hand-engineered to capture a specific high-level behavior. The internal formulation of a given policy can include a variety of local planning and control algorithms. We will cover different design choices for policies in the sections below. The key requirement for policy execution is that it works under forward simulation, which allows for a very broad class of algorithms. Thus, the per-agent joint density from Eq. 10.5 can now be approximated in terms of π_t^v:

$$p^v(x_t^v, x_{t+1}^v, z_t^v, a_t^v, \pi_t^v) = p(x_t^v)p(z_t^v|x_t^v)p(x_{t+1}^v|x_t^v, a_t^v)p(\pi_t^v|x_t, \mathbf{z_{1:t}})p(a_t^v|x_t^v, z_t^v, \pi_t^v). \tag{10.7}$$

Finally, since we have full authority over the policy executed by our controlled car $q \in V$, we can separate our agent from the other agents in $p(x_{t+1})$ as follows, using the per-agent distributions of Eq. 10.7:

---

[1]In this paper, we use the term *closed-loop policies* to mean policies that react to the presence of other agents, in a coupled manner. The same concept applies to the term *closed-loop forward simulation*.

$$p(x_{t+1}) \approx \iint_{X^q Z^q} p^q(x_t^q, x_{t+1}^q, z_t^q, a_t^q, \pi_t^q) \, dz_t^q \, dx_t^q$$

$$\prod_{v \in V | v \neq q} \left[\sum_{\Pi} \iint_{X^v Z^v} p^v(x_t^v, x_{t+1}^v, z_t^v, a_t^v, \pi_t^v) \, dz_t^v \, dx_t^v \right]. \quad (10.8)$$

We have thus far factored the action space from $p(x_{t+1})$ by assuming actions are given by the available policies.

However, Eq. 10.8 still requires integration over the state and observation spaces. We address this issue as follows. Given samples from $p(\pi_t^v | x_t, \mathbf{z}_{0:t})$ that assign a policy to each agent, we simulate forward both the robot and the other agents under their assigned policies to obtain sequences of predicted states and observations. These forward rollouts incorporate interactions. We evaluate the expected sum of rewards using these sample rollouts over the entire decision horizon in a computationally feasible manner.

We simplify the full POMDP solution in our approximate algorithm by reducing the decision to a limited set of policies and performing evaluations with a single set of policy assignments for each sample. The overall algorithm acts as a single-stage Markov decision process MDP, which does remove some scenarios from consideration, but for sufficiently high-level behaviors are not a major impediment to operation.

In addition, our approach approximates' policy outcomes as deterministic functions of state, but because policies internally incorporate closed-loop control, the actual outcomes of policies are well modeled by deterministic behavior. Even though we assume a deterministic transition model, we can incorporate uncertainty in terms of the state (see Sect. 10.5).

The policies used in this approach are still policies of the same form as in the POMDP literature, but under the constraint that the policy must be one of a predetermined policy set.

Algorithm 1 Policy selection.

Input:
- Current MAP estimate of the joint state, $x_0 \in X$.
- Set of available and applicable policies $\Pi' \subseteq \Pi$.
- Planning time horizon H.

1: $R \leftarrow \emptyset$ // Rewards for each forward propagation
2: **for each** $\pi \in \Pi'$ **do**
3:     $\Psi^\pi \leftarrow$ SIMULATEFORWARD(x_0, π, H) // Ψ^π captures all agents
4:     $R \leftarrow R \cup \{(\pi, \text{COMPUTEREWARD}(\Psi^\pi))\}$
5: **end for**
6: **return** $\pi^* \leftarrow$ SELECTBEST(R)   // As described in Eq. 10.3

Algorithm 1 describes the essence of the MPDM approach, where the functions SIMULATEFORWARD and COMPUTEREWARD have been expressed as a generic call

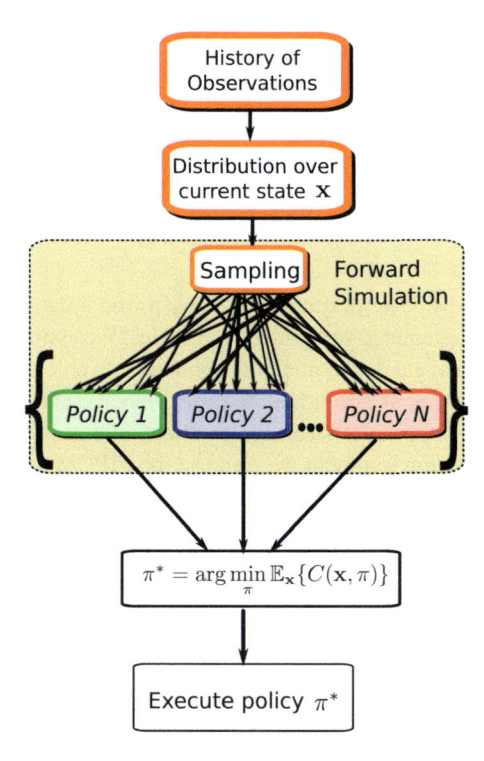

10 MPDM: Multi-policy Decision-Making from Autonomous Driving …

Fig. 10.2 Robot maintains a distribution over the state of each agent based on past observations. MPDM makes future predictions based on a motion model and the inferred states of all agents under consideration. For each policy π available and applicable to our robot, we simulate forward the system until the decision horizon H, which yields a set of simulated trajectories Ψ^π. We then evaluate the reward r_π for each rollout Ψ and finally select the policy π^* maximizing the expected reward. The number of samples is domain dependent, either drawn over the space of other vehicle policy assignments or over the space of the vehicle initial state (see Sects. 10.4 and 10.5). The process continuously repeats in a receding horizon manner. After one optimal policy π^* is chosen, then the system executes it

to adapt to domain-dependent characteristics. Fig. 10.2 is depicted an illustration of the MPDM algorithm in action.

In the remainder of the chapter, we present how we applied MPDM to two very different domains—autonomous driving and mobile indoor robots. We especially elaborate on the agent model we used for each—the handcrafted policies considered. The types of agents in the environment as well as the environmental constraints such as lanes in highways or walls/obstacles in indoor social settings affect the agent motion model used to make future predictions. The *agent motion model* governs the state that needs to be inferred based on historical observations. For instance, vehicles on a highway, probably given by the structure above mentioned, have a limited set of actions that they typically do, and abnormal behaviors are easily detected. On the other hand, pedestrians on a urban environment present a more complex modeling challenge, since more diverse action are possible. A vehicle moving on an unstructured environment, such as a parking lot, or a non-signaled area, results in more complex behaviors, which in turn require a more complex agent model.

10.4 Case Study 1: Autonomous Driving

The driving scenario is by design structured: in most traffic situations, vehicles behave in a regular, predictable manner, following traffic rules. We assume that such driving rules (lane rules, signals, etc.) determine the behavior of vehicles.

In our system, a state x_t^v is a tuple of the pose, velocity, and acceleration and an action a_t^v is a tuple of controls for steering, throttle, brake, shifter, and turn signals.

In this homogeneous environment, we can use the MPDM approach and assume that all vehicles are following a closed-loop policy at any given time—the car could be keeping to its lane, or changing lanes, or yielding. Thus, the policy takes on the additional role of a latent variable used to predict the future trajectories of the neighboring vehicles. The dynamics of a vehicle is determined by its present policy as well as the policies followed by all neighboring vehicles. The egovehicle maintains a posterior distribution over the latent variable values (closed-loop policies) based on prior observations.

In our system, we consider V as all vehicles that are tracked by our LIDAR system (typically within 50 m). An observation z_t^v, made by vehicle v, is a tuple including the observed poses and velocities of nearby vehicles and an occupancy grid of static obstacles. We consider only a limited field of view and do not account for observations that are far away from the robot.

Typically the position, velocity and acceleration of cars can be reliably tracked, but the uncertainty in the environment stems from the uncertainty about which closed-loop policy the other vehicles are following. We model uncertainty on the behavior of other agents with the following driver model:

$$D(x_t, z_t^v, a_t^v) = p(a_t^v | x_t, z_t^v), \quad (10.9)$$

where $a_t^v \in A$ is estimated as a switching sequence policies. The driver model $D(x_t, z_t^v, a_t^v)$ implicitly assumes that the instantaneous actions of each vehicle are independent of each other.

Changepoint detection can be used to efficiently infer when a vehicle changes its policy. Thus, MPDM is used for behavioral anticipation of other agents, inferring policies and then integrating anticipation with policy selection as described in Algorithm 1.

To carry out the evaluations, we use the autonomous vehicle platform (Fig. 10.3) for data collection and active autonomous driving. Our vehicle, a drive-by-wire Ford Fusion, is equipped with a sensor suite including four Velodyne HDL-32E 3D LIDAR scanners, an Applanix POS-LV 420 inertial navigation system (INS), and GPS. An

Fig. 10.3 Autonomous car platform. The vehicle is a Ford Fusion equipped with a sensor suite including four LIDAR units and survey-grade INS. All perception, planning, and control are performed onboard, from [1] © 2017 Springer, reproduced with permission

onboard five-node computer cluster performs all planning, control, and perception for the system in real time.

The vehicle uses prior maps of the area, and it operates on that capture information about the environment such as LIDAR reflectivity and road height and is used for localization and tracking of other agents. The road network is encoded as a metric-topological map that provides information about the location and connectivity of road segments, and lanes therein.

Estimates over the states of other traffic participants are provided by a dynamic object tracker running on the vehicle, which uses LIDAR range measurements. The geometry and location of static obstacles are also inferred onboard using LIDAR measurements.

10.4.1 Behavior Anticipation

Given the history of the agents in environment, our goal is to estimate a distribution over their current policy assignments so that we can sample over possible other vehicle policies during decision-making. The results presented in Fig. 10.4b indicate that we solve this in a robust manner.

The traffic-tracking dataset used to evaluate behavior anticipation consists of 67 dynamic object trajectories recorded in an urban area. Of these 67 trajectories, 18 correspond to "follow the lane" maneuvers and 20 to lane change maneuvers, recorded on a divided highway. The remaining 29 trajectories (shown in Fig. 10.4a) correspond to maneuvers observed at a four-way intersection regulated by stop signs.

(a) Collected dataset

(b) Precision and accuracy curves

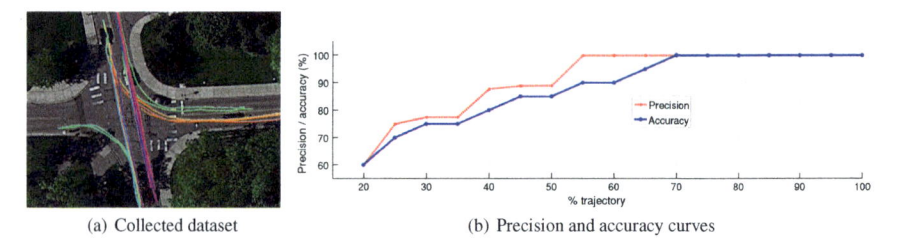

Fig. 10.4 A total of 29 trajectories in the traffic-tracking dataset used to evaluate our multi-policy framework, overlaid on satellite imagery (**a**). Precision and accuracy curves of current policy identification via changepoint detection evaluated at increasing subsequences of the trajectories. Our method provides over 85% accuracy and precision after only 50% of trajectory completion, while the closed-loop nature of our policies produce vehicle behavior that seeks safety in a timely manner regardless of anticipation performance (**b**). For more details on *accuracy* and *precision*; see [1]

All trajectories were recorded by the dynamic object tracker onboard the vehicle and extracted from approximately 3.5 h of total tracking data.

The performance of behavior anticipation can be observed in Fig. 10.4b. In all experiments, we use a C implementation of our system running on a single 2.8 GHz Intel i7 laptop computer. For a deeper presentation of the changepoint detection method for behavioral anticipation, and experiments please refer to [1, 4], and more on occlusions in [69].

10.4.2 Results

We tested the full behavioral anticipation and decision-making system in both real-world and simulated highway traffic scenarios to demonstrate feasibility in a real vehicle environment and evaluate the effect of policy sampling strategies on decision results. The two-vehicle scenario we used is illustrated in Fig. 10.5, showing both our initial simulation of the test scenario and the real-world driving case.

In particular, this scenario highlights a case where identifying the behavior of another vehicle, in this case the second lane change of vehicle 2, causes the system to decide to initiate our lane change as soon as the it is clear the vehicle 2 is going to leave the lane. This extends our previous experimental results from [3], which demonstrated many trials of simple overtaking of a vehicle on a two-lane road assuming a single possible behavior for the passed vehicle.

In both real-world and simulated cases, we ran Algorithm 1 using a 0.25 s simulation step with a 10 s rollout horizon, with the same multi-threaded implementation of policy selection. The target execution rate for policy selection is 1 Hz, with a separate thread for executing the current policy running at 30 Hz. The process uses four threads for sample evaluation, and because the samples are independent, the speedup

10 MPDM: Multi-policy Decision-Making from Autonomous Driving ... 213

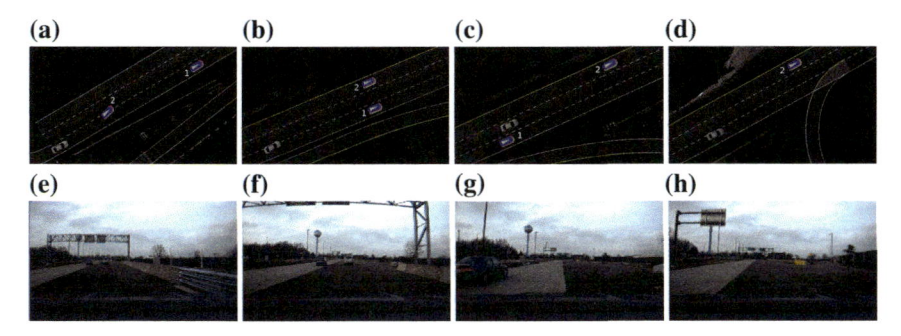

Fig. 10.5 Two-vehicle passing scenario executed in both simulation (top) and on our test vehicle, shown from the forward-facing camera. Note while the vehicles do not have the same timing in both cases, the structure of the scenario is the same in both. In this scenario, the egovehicle starts behind both traffic vehicles in the right lane of the three-lane road. The traffic vehicle 1 drives in the right lane along the length of the road, while traffic vehicle 2 makes two successive lane changes to the left. We remain in the right lane behind vehicle 1 until vehicle 2 initiates a lane change from the center to left lane, and at that point, we make a lane change to the center lane. We pass both vehicles and return to the right lane, from [1] © 2017 Springer, reproduced with permission

from multi-threading is roughly linear so long as all threads are kept busy. In this scenario, for both the egovehicle and the traffic vehicles, we used a pool of three policies that are representative of highway environments:

$$\Pi = \{\text{lane-nominal, lane-change-left, lane-change-right}\}.$$

In the context of autonomous cars, our typical metrics capture accomplishment of goals, safety, implementation of "soft" driving rules, and rider comfort.

We use a straightforward set of metrics in this scenario to compose the reward function with empirically tuned weights. The metrics used are as follows:

1. *Distance to goal*: scores how close the final rollout pose is to the goal.
2. *Lane bias*: penalizes being far from the right lane.
3. *Maximum yaw rate*: penalizes abrupt steering.
4. *Dead-end distance*: penalizes staying in dead-end lanes depending on distance to the end.

These costs are designed to approximate the idealized value function that might come from a classical POMDP solution and to avoid biases due to heuristic cost functions.

As can be seen through the policy reward trends in Fig. 10.6, there are clear decision points in which we choose to execute a new policy, which results in stable policy selection decisions. Discontinuities, such as the reward for *lane-change-right*, are expected as some policies are applicable less often, and in the middle of a maneuver such as a lane change, it is not possible that no policies can be initiated. In cases where a policy cannot be preempted until completed, such as lane-changes, another policy may have a higher reward but not induce policy switch due to concurrent

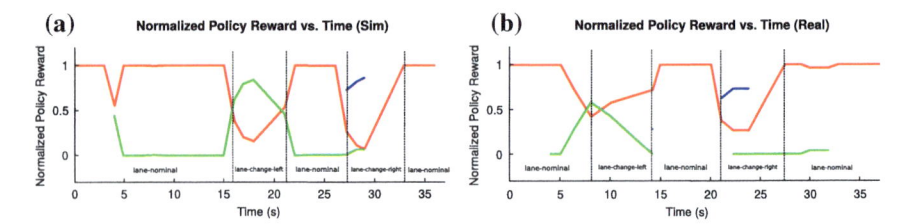

Fig. 10.6 These time-series plots show rewards for each policy (where policies *lane-nominal*, *lane-change-left* and *lane-change-right* are red, green, and blue, respectively) is available to the egovehicle for both the simulated (**a**) and real-world version of the test scenario (**b**), with policy rewards normalized at each timestep. The dashed lines indicate the transitions between currently running policies based on the result of the elections. Discontinuities are due to a policy not being applicable, for reasons such as a vehicle blocking a lane change, or *lane-change-right* not being feasible from the right lane, from [1] © 2017 Springer, reproduced with permission

policy execution and selection, such as in Fig. 10.6b at 10 s, where we continue a lane-change even though *lane-nominal* has a locally higher reward. The reward in this case is higher because trajectory generation within the lane-change policy expects to start at a lane center, not while between lanes as during the lane change itself.

From the demonstrations in both simulation and real-world experiments, the policy selection process makes qualitatively reasonable decisions as expected given the reward metric weights. Further, evaluation of the correctness of decisions made, however, will require larger-scale testing with real-world traffic in order to determine whether decisions made are statistically consistent with the median human driver.

10.5 Case Study 2: Social Environment

We use MPDM on an indoor and social environment, where our robot navigates among pedestrians. In this environment, outcomes are harder to predict, such as people suddenly stopping or changing directions, whereas in the driving scenario, other vehicles generate less unexpected events and that allowed us to define a more complex and accurate model of vehicles.

The agent model must capture the dynamics of agents including reaction to other agents. At the same time, agents (humans) can instantaneously stop or change direction without signaling making it very difficult to predict future scenarios. The choice of model trades off accuracy with computational efficiency. Complex models capturing interactions between groups of pedestrians and crowd dynamics may be more accurate but inferring parameters may be computationally expensive and may require observations that are not feasible under the provided sensor setup.

The key aspects of more unstructured environments are that inference over the agent's state is subject to a great inaccuracy. In order to apply the MPDM approach to its full potential, we propose a computationally light agent model and inference procedure. In consequence, the robot can re-plan frequently, which helps reduce the impact of this uncertainty. We use a simple reactive motion model. Each pedestrian in the vicinity treats all other agents as obstacles and uses a potential field based on the social force model (SFM) [50, 54] to guide it toward its goal with a desired speed. The forward rollouts capture the interactions between agents. For each pedestrian, the goal is not directly observable. It is assumed to be one of a small set of salient points and is estimated using a naive Bayes classifier. The significant parameters of this model that typically contribute most to the uncertainty are the inferred goal and the preferred speed of the agent. Hence, MPDM maintains a distribution over the current state of each agent and samples initial configurations. Each sample is then forward simulated and contributes toward the expected utility of the policy.

Dynamically switching between the candidate policies allows the robot to adapt to different situations. In the social environment, the set of most representative policies are:

$$\Pi = \{\text{Go-solo, Follow other agent, Stop}\}.$$

A robot executing the *Go-Solo* policy treats all other agents as obstacles and uses a potential field based on the social force model (SFM) [50, 54] to guide it toward its goal.

In addition to the *Go-Solo* and the *Stop* policy, the robot can use the *Follow* policy to deal with certain situations. Our intuition is that in a crowd, the robot may choose to *Follow* another person sacrificing speed but delegating the task of finding a path to a human. *Following* could also be more suitable than overtaking a person in a cluttered scenario as it allows the robot to progress toward its goal without inconveniencing other pedestrians.

We show the benefits of switching between multiple policies in terms of navigation performance, quantified by metrics for progress made, and inconvenience to fellow agents. We demonstrate the robustness of MPDM to measurement uncertainty (Sect. 10.5.1). Finally, we test the MPDM on a real environment and evaluate the results (Sect. 10.5.1.1).

Evaluating navigation behavior objectively is a challenging task and unfortunately, there are no standard metrics.

We propose three metrics that quantify different aspects of the emergent navigation behavior of the robot (for more details, please refer to [2]).

1. *Progress* (PG) - measures distance made good.
2. *Force* (F) - penalizes close encounters with other agents, calculated at each time step.
3. *Blame* (B) - penalizes velocity at the time of close encounters which is not captured by *Force*.

In order for the robot's emergent behavior to be socially acceptable, each policy's utility is estimated trading-off the distance travelled toward the goal (*Progress*) with the potential disturbance caused to fellow agents (*Force*).

10.5.1 Simulation

We simulate an indoor domain, freely traversed by a set of agents while the robot tries to reach a goal. We use the Intel i7 processor and 8GB RAM for our simulator and LCM [70] for inter-process communication.

We assume that the position of the robot, agents, the goal point, and obstacles are known in some local coordinate system. However, the accuracy of motion predictions is improved by knowing more about the structure of the building since the position of walls and obstacles can influence the behavior of other agents over the 3 s planning horizon. Our implementation achieves these through a global localization system with a known map, but our approach could be applied more generally.

The hallway domain (Fig. 10.7) is modeled on a 3 m $\times$ 25 m hallway at the University of Michigan. The maximum permitted acceleration is 3 m/s^2 while the maximum speed $|v|_{max}$ is set to 0.8 m/s. MPDM is carried out at 3 Hz to match the frequency of the sensing pipeline for state estimation in the real-world experiment. The planning horizon is 3 s into the future.

10.5.1.1 Simulations with Noise

For MPDM, the more accurate our model of the dynamic agents, the better is the accuracy of the predicted joint states. Most models of human motion, especially in complicated situations, fail to predict human behavior accurately. This motivates us to extensively test how robust our approach is to noisy environments.

In our simulator, the observations z are modeled using a stationary Gaussian distribution with uncorrelated variables for position, speed, and orientation for the agent. We parameterize this uncertainty by a scale factor k_z: $\{\sigma_{p_x}, \sigma_{p_y}, \sigma_{|v|}, \sigma_\theta\} = k_z \times \{2\,\text{cm}, 2\,\text{cm}, 2\,\text{cm/s}, 3°\}$. The corresponding diagonal covariance matrix is denoted by $\text{diag}(\sigma_{p_x}, \sigma_{p_y}, \sigma_{|v|}, \sigma_\theta)$. These uncertainties are propagated in the posterior estate estimation $P(x|z)$.

Fig. 10.7 Simulated indoor domain chosen to study our approach. The hallway domain where 15 agents are let loose with the robot, and they patrol the hallway while the robot tries to reach its destination, from [2] © 2016 IEEE, reproduced with permission

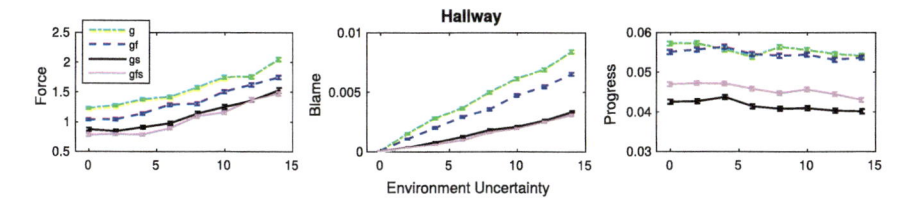

Fig. 10.8 Simulation results varying uncertainty in the environment (k_z) for a fixed posterior uncertainty (k_e). We show results for four combinations of the policies, varying the flexibility of MPDM: *Go-Solo* (g), *Go-Solo* and *Follow* (gf), *Go-Solo* and *Stop* (gs), and the full policy set (gfs). The data is averaged in groups of 10. We show the mean and standard error. *Left* Increasing the noise in the environment makes the robot more susceptible to disturbing other agents and vice versa. We can observe that the *Force* when combining all the policies (gfs) is much lower than when using a single policy (g) in the hallway domain. *Center* A lower *Blame* indicates better behavior as the robot is less often the cause of inconvenience. The robustness of MPDM can be observed in milder slope across both domains. *Right* Higher *Progress* is better. The *Go-Solo* performs better, however at the price of being much worse in *Force* and *Blame*. With more flexibility, (gfs) is able to achieve greater *Progress* and lower *Force* as compared to (gf), from [2] © 2016 IEEE, reproduced with permission

The robot's estimator makes assumptions about the observation noise which may or may not match the noise injected by the simulator. This can lead to over and underconfidence which affects decision-making. In this section, we explore the robustness of the system in the presence of these types of errors. We define the assumed uncertainty by the estimator through a scale factor k_e, exactly as described above.

Varying the environment uncertainty k_z for a fixed level of estimator uncertainty k_e to understand how MPDM performs. We have studied the impact of different levels of environment uncertainty (k_z) at regular intervals of diag(4 cm, 4 cm, 4 cm/s, 6°). The estimation uncertainty k_e is fixed at diag(10 cm, 10 cm, 10 cm/s, 15°).

Figure 10.8 shows the performance of the robot for the hallway domain. We observe that the *Blame* increases at a lowest rate for MPDM with the complete policy set. If the option of stopping is removed, we notice that the addition of the follow policy allows the robot to maintain comparable *Progress* while reducing the force and *Blame* associated. Given the option of stopping, the robot still benefits from the option of following as it can make more *Progress* while keeping *Blame* and *Force* lower.

We observe that MPDM allows the robot to maintain *Progress* toward the goal while exerting less *Force* and incurring less *Blame*. We also observe that the robot is more robust to noise in terms of *Blame* incurred (lesser rate of increase).

10.5.2 Real-World Experiments

Our real-world experiments have been carried out in the hallway that the simulated hallway domain was modeled on Sect. 10.5.1. We implemented our system on the MAGIC robot [71], a differential drive platform equipped with a Velodyne 16 laser scanner used for tracking and localization. An LED grid mounted on the head of the robot has been used to visually indicate the policy chosen at any time.

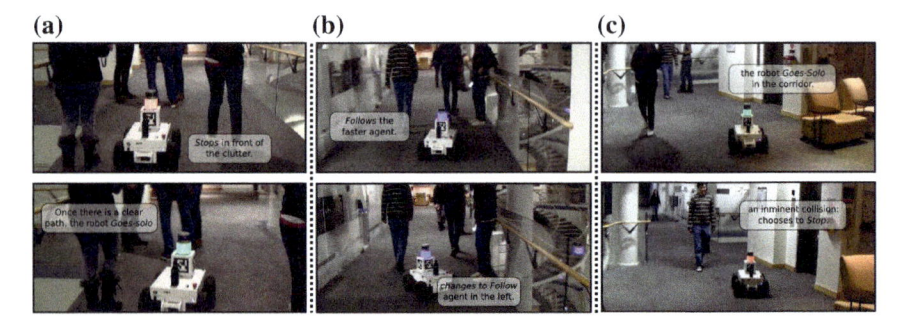

Fig. 10.9 Real situations (**a–c**) illustrating the nature of the MPDM. On the top row is depicted some situations while testing the robot navigation in a real environment. On the bottom row are shown the same configurations, but delayed by a few seconds. The lights on the robot indicate the policy being executed, being green for *Go-Solo*, blue *Follow*, and red *Stop*. By dynamically switching between policies, the robot can deal with a variety of situations, from [2] © 2016 IEEE, reproduced with permission

Fig. 10.10 Mean and standard error for the performance metrics over 10 s intervals form real-world experiments. All measures are normalized based on the corresponding mean value for the *Go-Solo* policy. MPDM shows much better *Force* and *Blame* costs than only *Go-Solo* at the price of slightly reducing its *Progress*, from [2] © 2016 IEEE, reproduced with permission

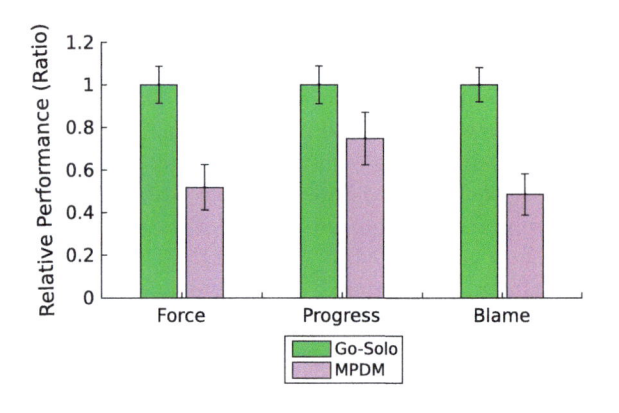

During two days of testing, a group of 8 volunteers was asked to patrol the hallway, given random initial and goal positions, similar to the simulation results proposed in Sect. 10.5.1. The robot alternated between using MPDM and using the *Go-Solo* policy exclusively every five minutes. The performance metrics were recorded every second, constituting a total of 4.8k measurements.

Figure 10.9 depicts some of the challenging situations that our approach has tackled successfully. On the *Right* and *Left* scenes, the robot chooses to *Stop* avoiding the "freezing robot behavior" which would result in high values of *Blame* and *Force*. As soon as the dynamic obstacles are no longer an hindrance, the robot changes the policy to execute and *Goes-Solo*. In Fig. 10.9-*Center*, we show an example of the robot executing the *Follow* policy, switching between leaders in order to avoid inconveniencing the person standing by the wall. The video provided[2] clearly shows the limitations of the *Go-Solo* and how MPDM solves these limitations.

---

[2]https://April.eecs.umich.edu/media/mehta2016iros.mp4.

Figure 10.10 shows the results of MPDM compared to a constant navigation policy - *Go-Solo*. We show that our observations based on simulations hold in real environments. Specifically, MPDM performs much better, roughly 50%, in terms of *Force* and *Blame* while sacrificing roughly 30% in terms of *Progress*. This results in the more desirable behavior for navigation in social environments that is qualitatively evident in the video provided.

10.6 Conclusion

We have introduced a principled framework for decision-making in environments under uncertainty with extensively coupled interactions between agents as an approximate POMDP solver. By explicitly modeling reasonable behaviors of both our system and other agents' policies, we make informed high-level behavioral decisions that account for the consequences of our actions. This framework has also applications to ADAS and ADF systems: MPDM evaluates efficiently and robustly a set policies, which might be obtained directly from drivers' data or adjusted to every individual driver.

In this chapter, we have also presented two cases, an autonomous car driving and a robot navigation. MPDM has been successfully applied to each of these very different cases by carefully taking different assumptions. In the case of autonomous driving, policies are low-level maneuvers where complex interactions take place. We therefore predict intentions over a longer history as well as allow a longer planning budget. These policies correspond to typical drivers' maneuvers, so an intermediate step for driving assistance would perfectly suit MPDM. While navigating in indoor social environments, in order to compensate for the inaccuracies in the prediction model, we required a faster update of the policy selection, but tolerant to higher levels of uncertainty. As we have shown, this approach is feasible in real-world test cases and can be implemented online as it is required for autonomous driving and robot navigation in real environments.

Acknowledgements This work was supported by a grant from Ford Motor Company via the Ford-UM Alliance under award N015392, DARPA YIP grant under award D13AP00059, CyberSEES grant award 1442773, and ARIA (TRI) grant award N021563.

Parts of this work have been previously published in [1] which is under Copyright by Springer, 2017. These parts are reused with the permission of Springer which is acknowledged with high appreciation.

References

1. Galceran, E., Cunningham, A.G., Eustice, R.M., Olson, E.: Multipolicy decision-making for autonomous driving via changepoint-based behavior prediction: theory and experiment. In: Autonomous Robots, pp. 1–16 (2017)

2. Mehta, D., Ferrer, G., Olson, E.: Autonomous navigation in dynamic social environments using multi-policy decision making. In: Proceedings of the IEEE/RSJ International Conference on Intelligent Robots and Systems, pp. 1190–1197 (2016)
3. Cunningham, A.G., Galceran, E., Eustice, R.M., Olson, E.: MPDM: multipolicy decision-making in dynamic, uncertain environments for autonomous driving. In: Proceedings of the IEEE International Conference on Robotics and Automation. Seattle, WA, USA (2015)
4. Galceran, E., Cunningham, A.G., Eustice, R.M., Olson, E.: Multipolicy decision-making for autonomous driving via changepoint-based behavior prediction. In: Proceedings of the Robotics: Science and Systems Conference. Rome, Italy (2015)
5. Choi, J., Eoh, G., Kim, J., Yoon, Y., Park, J., Lee, B.H.: Analytic collision anticipation technology considering agents' future behavior. In: Proceedings of the IEEE/RSJ International Conference on Intelligent Robots and Systems, pp. 1656–1661. Taipei, Taiwan (2010)
6. Ohki, T., Nagatani, K., Yoshida, K.: Collision avoidance method for mobile robot considering motion and personal spaces of evacuees. In: Proceedings of the IEEE/RSJ International Conference on Intelligent Robots and Systems, pp. 1819–1824. Taipei, Taiwan (2010)
7. Petti, S., Fraichard, T.: Safe motion planning in dynamic environments. In: Proceedings of the IEEE/RSJ International Conference on Intelligent Robots and Systems, pp. 2210–2215. Edmonton, AB, Canada (2005)
8. Du Toit, N., Burdick, J.: Robotic motion planning in dynamic, cluttered, uncertain environments. In: Proceedings of the IEEE International Conference on Robotics and Automation, pp. 966–973. Anchorage, AK, USA (2010)
9. Fulgenzi, C., Tay, C., Spalanzani, A., Laugier, C.: Probabilistic navigation in dynamic environment using rapidly-exploring random trees and Gaussian processes. In: Proceedings of the IEEE/RSJ International Conference on Intelligent Robots and Systems, pp. 1056–1062. Nice, France (2008)
10. Du Toit, N.E., Burdick, J.W.: Robot motion planning in dynamic, uncertain environments. IEEE Trans. Robot. **28**(1), 101–115 (2012)
11. Havlak, F., Campbell, M.: Discrete and continuous, probabilistic anticipation for autonomous robots in urban environments. IEEE Trans. Robot. **30**(2), 461–474 (2014)
12. Dagli, I., Brost, M., Breuel, G.: Agent technologies, infrastructures, tools, and applications for e-services. NODe 2002 Agent-Related Workshops. Chapter Action Recognition and Prediction for Driver Assistance Systems Using Dynamic Belief Networks, pp. 179–194. Springer, Berlin, Heidelberg (2003)
13. Gindele, T., Brechtel, S., Dillmann, R.: Learning driver behavior models from traffic observations for decision making and planning. In: IEEE Intelligent Transportation Systems Magazine, pp. 69–79 (2015)
14. Broadhurst, A., Baker, S., Kanade, T.: Monte Carlo road safety reasoning. In: Proceedings of the IEEE Intelligent Vehicles Symposium, pp. 319–324. Las Vegas, NV, USA (2005)
15. Ferguson, D., Darms, M., Urmson, C., Kolski, S.: Detection, prediction, and avoidance of dynamic obstacles in urban environments. In: Proceedings of the IEEE Intelligent Vehicles Symposium, pp. 1149–1154. Eindhoven, Netherlands (2008)
16. Hardy, J., Campbell, M.: Contingency planning over probabilistic obstacle predictions for autonomous road vehicles. IEEE Trans. Robot. **29**(4), 913–929 (2013)
17. Joseph, J., Doshi-Velez, F., Huang, A.S., Roy, N.: A Bayesian nonparametric approach to modeling motion patterns. Auton. Robots **31**(4), 383–400 (2011)
18. Kim, K., Lee, D., Essa, I.: Gaussian process regression flow for analysis of motion trajectories. In: Proceedings of the IEEE International Conference on Computer Vision, pp. 1164–1171. Barcelona, Spain (2011)
19. Trautman, P., Krause, A.: Unfreezing the robot: Navigation in dense, interacting crowds. In: Proceedings of the IEEE/RSJ International Conference on Intelligent Robots and Systems, pp. 797–803. Taipei, Taiwan (2010)
20. Aoude, G.S., Luders, B.D., Joseph, J.M., Roy, N., How, J.P.: Probabilistically safe motion planning to avoid dynamic obstacles with uncertain motion patterns. Auton. Robots **35**(1), 51–76 (2013)

21. Tran, Q., Firl, J.: Modelling of traffic situations at urban intersections with probabilistic non-parametric regression. In: Proceedings of the IEEE Intelligent Vehicles Symposium, pp. 334–339. Gold Coast City, Australia (2013)
22. Tran, Q., Firl, J.: Online maneuver recognition and multimodal trajectory prediction for intersection assistance using non-parametric regression. In: Proceedings of the IEEE Intelligent Vehicles Symposium, pp. 918–923. Dearborn, MI, USA (2014)
23. Kuderer, M., Gulati, S., Burgard, W.: Learning driving styles for autonomous vehicles from demonstration. In: Proceedings of the IEEE International Conference on Robotics and Automation, pp. 2641–2646 (2015)
24. DARPA: DARPA Urban Challenge. http://archive.darpa.mil/grandchallenge/ (2007)
25. Montemerlo, M., et al.: Junior: the Stanford entry in the urban challenge. J. Field Robot. **25**(9), 569–597 (2008)
26. Miller, I., et al.: Team Cornell's skynet: robust perception and planning in an urban environment. J. Field Robot. **25**(8), 493–527 (2008)
27. Urmson, C., Anhalt, J., Bagnell, D., Baker, C., Bittner, R., Clark, M.N., Dolan, J., Duggins, D., Galatali, T., Geyer, C., Gittleman, M., Harbaugh, S., Hebert, M., Howard, T.M., Kolski, S., Kelly, A., Likhachev, M., McNaughton, M., Miller, N., Peterson, K., Pilnick, B., Rajkumar, R., Rybski, P., Salesky, B., Seo, Y.W., Singh, S., Snider, J., Stentz, A., Whittaker, W., Wolkowicki, Z., Ziglar, J., Bae, H., Brown, T., Demitrish, D., Litkouhi, B., Nickolaou, J., Sadekar, V., Zhang, W., Struble, J., Taylor, M., Darms, M., Ferguson, D.: Autonomous driving in urban environments: boss and the Urban Challenge. J. Field Robot. **25**(8), 425–466 (2008)
28. Ferguson, D., Howard, T.M., Likhachev, M.: Motion planning in urban environments. J. Field Robot. **25**(11–12), 939–960 (2008)
29. Werling, M., Ziegler, J., Kammel, S., Thrun, S.: Optimal trajectory generation for dynamic street scenarios in a frenet frame. In: Proceedings of the IEEE International Conference on Robotics and Automation, pp. 987–993. Anchorage, AK, USA (2010)
30. Xu, W., Wei, J., Dolan, J., Zhao, H., Zha, H.: A real-time motion planner with trajectory optimization for autonomous vehicles. In: Proceedings of the IEEE International Conference on Robotics and Automation, pp. 2061–2067. Saint Paul, MN, USA (2012)
31. Bai, H., Hsu, D., Lee, W.S.: Integrated perception and planning in the continuous space: a POMDP approach. Int. J. Robot. Res. **33**(9), 1288–1302 (2014)
32. Kurniawati, H., Hsu, D., Lee, W.: SARSOP: Efficient point-based POMDP planning by approximating optimally reachable belief spaces. In: Proceedings of the Robotics: Science and Systems Conference. Zurich, Switzerland (2008)
33. Silver, D., Veness, J.: Monte-Carlo planning in large POMDPs. In: Lafferty J., Williams, C., Shawe-Taylor, J., Zemel, R., Culotta, A. (eds.) Advances in Neural Information Processing Systems, vol. 23, pp. 2164–2172. Curran Associates, Inc. (2010)
34. Madani, O., Hanks, S., Condon, A.: On the undecidability of probabilistic planning and related stochastic optimization problems. Artif. Intell. **147**(1–2), 5–34 (2003)
35. Papadimitriou, C.H., Tsitsiklis, J.N.: The complexity of Markov decision processes. Math. Oper. Res. **12**(3), 441–450 (1987)
36. Thrun, S.: Monte Carlo POMDPs. In: Proceedings of the Advances in Neural Information Processing Systems Conference pp. 1064–1070 (2000)
37. Candido, S., Davidson, J., Hutchinson, S.: Exploiting domain knowledge in planning for uncertain robot systems modeled as pomdps. In: Proceedings of the IEEE International Conference on Robotics and Automation, pp. 3596–3603. Anchorage, AK, USA (2010)
38. Lee, T., Kim, Y.J.: Massively parallel motion planning algorithms under uncertainty using POMDP. Int. J. Robot. Res. **35**(8), 928–942 (2016)
39. Wei, J., Dolan, J.M., Snider, J.M., Litkouhi, B.: A point-based MDP for robust single-lane autonomous driving behavior under uncertainties. In: Proceedings of the IEEE International Conference on Robotics and Automation, pp. 2586–2592. Shanghai, China (2011)
40. Ulbrich, S., Maurer, M.: Probabilistic online pomdp decision making for lane changes in fully automated driving. In: Proceedings of the IEEE Intelligent Transportation Systems Conference, pp. 2063–2067 (2013)

41. Brechtel, S., Gindele, T., Dillmann, R.: Probabilistic MDP-behavior planning for cars. In: Proceedings of the IEEE Intelligent Transportation Systems Conference, pp. 1537–1542 (2011)
42. Bandyopadhyay, T., Jie, C.Z., Hsu, D., Ang, M.H., Rus, D., Frazzoli, E.: In: Experimental Robotics: The 13th International Symposium on Experimental Robotics, Chapter Intention-Aware Pedestrian Avoidance, pp. 963–977. Springer (2013)
43. Brechtel, S., Gindele, T., Dillmann, R.: Probabilistic decision-making under uncertainty for autonomous driving using continuous POMDPs. In: Proceedings of the IEEE Intelligent Transportation Systems Conference, pp. 392–399 (2014)
44. Bandyopadhyay, T., Won, K., Frazzoli, E., Hsu, D., Lee, W., Rus, D.: Intention-aware motion planning. In: Frazzoli, E., Lozano-Perez, T., Roy, N., Rus, D. (eds.) In: Proceedings of the International Workshop on the Algorithmic Foundations of Robotics. Springer Tracts in Advanced Robotics, vol. 86, pp. 475–491. Springer, Berlin, Heidelberg (2013)
45. He, R., Brunskill, E., Roy, N.: Efficient planning under uncertainty with macro-actions. J. Artif. Intell. Res. **40**, 523–570 (2011)
46. Somani, A., Ye, N., Hsu, D., Lee, W.S.: DESPOT: Online POMDP planning with regularization. In: Burges, C., Bottou, L., Welling, M., Ghahramani, Z., Weinberger, K. (eds.) Advances in Neural Information Processing Systems, vol. 26, pp. 1772–1780. Curran Associates, Inc. (2013)
47. van den Berg, J., Guy, S.J., Lin, M., Manocha, D.: Reciprocal n-body collision avoidance. Robotics Research, Springer Tracts in Advanced Robotics **70**, 3–19 (2011)
48. Guy, S.J., Chhugani, J., Kim, C., Satish, N., Lin, M., Manocha, D., Dubey, P.: Clearpath: highly parallel collision avoidance for multi-agent simulation. In: Proceedings of the 2009 ACM SIGGRAPH/Eurographics Symposium on Computer Animation, pp. 177–187. ACM (2009)
49. Khatib, O.: Real-time obstacle avoidance for manipulators and mobile robots. Int. J. Robot. Res. **5**(1), 90–98 (1986)
50. Ferrer, G., Garrell, A., Sanfeliu, A.: Social-aware robot navigation in urban environments. In: European Conference on Mobile Robotics, pp. 331–336 (2013)
51. Sisbot, E.A., Marin-Urias, L.F., Alami, R., Simeon, T.: A human aware mobile robot motion planner. IEEE Trans. Robot. **23**(5), 874–883 (2007)
52. Svenstrup, M., Bak, T., Andersen, H.J.: Trajectory planning for robots in dynamic human environments. In: Proceedings of the IEEE/RSJ International Conference on Intelligent Robots and Systems, pp. 4293–4298 (2010)
53. Huang, W.H., Fajen, B.R., Fink, J.R., Warren, W.H.: Visual navigation and obstacle avoidance using a steering potential function. Robot. Auton. Syst. **54**(4), 288–299 (2006)
54. Helbing, D., Molnár, P.: Social force model for pedestrian dynamics. Phys. Rev. E **51**(5), 4282 (1995)
55. Koren, Y., Borenstein, J.: Potential field methods and their inherent limitations for mobile robot navigation. In: Proceedings of the IEEE International Conference on Robotics and Automation, pp. 1398–1404 (1991)
56. Brock, O., Khatib, O.: High-speed navigation using the global dynamic window approach. Proceedings of the IEEE International Conference on Robotics and Automation **1**, 341–346 (1999)
57. Kretzschmar, H., Spies, M., Sprunk, C., Burgard, W.: Socially compliant mobile robot navigation via inverse reinforcement learning. Int. J. Robot. Res. (2016)
58. Kuderer, M., Kretzschmar, H., Sprunk, C., Burgard, W.: Feature-based prediction of trajectories for socially compliant navigation. In: Proceedings of Robotics: Science and Systems (RSS) (2012)
59. Luber, M., Spinello, L., Silva, J., Arras, K.O.: Socially-aware robot navigation: a learning approach. In: Proceedings of the IEEE/RSJ International Conference on Intelligent Robots and Systems, pp. 902–907 (2012)
60. Ziebart, B.D., Ratliff, N., Gallagher, G., Mertz, C., Peterson, K., Bagnell, J.A., Hebert, M., Dey, A.K., Srinivasa, S.: Planning-based prediction for pedestrians. In: Proceedings of the IEEE/RSJ International Conference on Intelligent Robots and Systems, pp. 3931–3936 (2009)

61. Ferrer, G., Garrell, A., Herrero, F., Sanfeliu, A.: Robot social-aware navigation framework to accompany people walking side-by-side. In: Autonomous Robots, pp. 1–19 (2016)
62. Kuderer, M., Burgard, W.: An approach to socially compliant leader following for mobile robots. In: International Conference on Social Robotics, pp. 239–248. Springer (2014)
63. Stein, P., Spalanzani, A., Santos, V., Laugier, C.: Leader following: a study on classification and selection. Robot. Auton. Syst. **75**(Part A), 79 – 95 (2016)
64. Ferrer, G.: Social robot navigation in urban dynamic environments. Ph.D. thesis, Universitat Politèctnica de Catalunya, Spain (October 2015)
65. Ferrer, G., Sanfeliu, A.: Multi-objective cost-to-go functions on robot navigation in dynamic environments. In: Proceedings of the IEEE/RSJ International Conference on Intelligent Robots and Systems, pp. 3824–3829 (2015)
66. Fulgenzi, C., Spalanzani, A., Laugier, C.: Probabilistic motion planning among moving obstacles following typical motion patterns. In: Proceedings of the IEEE/RSJ International Conference on Intelligent Robots and Systems, pp. 4027–4033. IEEE (2009)
67. Trautman, P., Ma, J., Murray, R.M., Krause, A.: Robot navigation in dense human crowds: Statistical models and experimental studies of human-robot cooperation. Int. J. Robot. Res. **34**(3), 335–356 (2015)
68. Foka, A., Trahanias, P.: Probabilistic Autonomous Robot Navigation in Dynamic Environments with Human Motion Prediction. Int. J. Soc. Robot. **2**(1), 79–94 (2010). https://doi.org/10.1007/s12369-009-0037-z
69. Galceran, E., Olson, E., Eustice, R.M.: Augmented vehicle tracking under occlusions for decision-making in autonomous driving. In: Proceedings of the IEEE/RSJ International Conference on Intelligent Robots and Systems, pp. 3559–3565. Hamburg, Germany (2015)
70. Huang, A.S., Olson, E., Moore, D.C.: LCM: lightweight communications and marshalling. In: Proceedings of the IEEE/RSJ International Conference on Intelligent Robots and Systems, pp. 4057–4062 (2010)
71. Olson, E., Strom, J., Morton, R., Richardson, A., Ranganathan, P., Goeddel, R., Bulic, M., Crossman, J., Marinier, B.: Progress toward multi-robot reconnaissance and the magic 2010 competition. J. Field Robot. **29**(5), 762–792 (2012)